George Washington Crile

Experimental Research into the Surgery of the Respiratory System

George Washington Crile

Experimental Research into the Surgery of the Respiratory System

ISBN/EAN: 9783337181406

Printed in Europe, USA, Canada, Australia, Japan

Cover: Foto ©berggeist007 / pixelio.de

More available books at **www.hansebooks.com**

EXPERIMENTAL RESEARCH

INTO THE

SURGERY OF THE RESPIRATORY SYSTEM

AN ESSAY AWARDED THE NICHOLAS SENN PRIZE
BY THE AMERICAN MEDICAL ASSO-
CIATION FOR 1898

BY

GEORGE W. CRILE, A.M., M.D., Ph.D.

PROFESSOR OF THE PRINCIPLES OF SURGERY AND APPLIED ANATOMY IN THE CLEVELAND COLLEGE OF
PHYSICIANS AND SURGEONS; FORMERLY PROFESSOR OF PHYSIOLOGY IN THE MEDICAL
DEPARTMENT OF THE UNIVERSITY OF WOOSTER; ATTENDING SURGEON TO
THE ST. ALEXIS AND THE CLEVELAND GENERAL HOSPITALS

PHILADELPHIA
J. B. LIPPINCOTT COMPANY

COPYRIGHT, 1899
BY
GEORGE W. CRILE

PRINTED BY J. B. LIPPINCOTT COMPANY, PHILADELPHIA, U. S. A.

TABLE OF CONTENTS

	PAGE
I.—INTRODUCTION	7
II.—METHODS OF RESEARCH AND NOTATION	8
III.—ON THE CAUSE OF THE PHENOMENA ATTENDING THE INHALATION OF HOT AIR AND FLAME	13
1. Preliminary Remarks	13
2. Protocols	14
3. Summary of Experimental Evidence	16
IV.—ON THE EFFECT OF FILLING THE CHEST WITH FLUID	18
1. Preliminary Remarks	18
2. Protocols	18
3. Summary of Experimental Evidence	20
V.—ON THE EFFECT OF PROLONGED MANIPULATION OF THE BRACHIAL PLEXUS AND THE NERVES SUPPLYING SOME OF THE MUSCLES OF RESPIRATION	21
1. Preliminary Remarks	21
2. Protocols	21
3. Summary	24
4. Practical Application	25
VI.—ON THE CAUSE OF COLLAPSE OR DEATH FROM BLOWS UPON THE LOWER CHEST AND EPIGASTRIUM	26
1. Preliminary Remarks	26
2. Protocols	26
3. Summary of Experimental Evidence	36
VII.—ON THE MECHANISM OF DROWNING	39
1. Preliminary Remarks	39
2. Protocols	41
3. Summary of Experimental Evidence	58
VIII.—ON THE CAUSE OF CERTAIN SYMPTOMS OBSERVED ON ENTERING AN ATMOSPHERE OF INCREASED BAROMETRIC PRESSURE	66
1. Preliminary Remarks	66
2. Protocols	66
3. Summary of Experimental Evidence	69

TABLE OF CONTENTS

	PAGE
IX.—ON FOREIGN BODIES IN THE PHARYNX AND ŒSOPHAGUS	70
1. Preliminary Remarks	70
2. Protocols	70
3. Summary of Experimental Evidence	77
4. Some Observations	79
X.—ON FOREIGN BODIES IN THE TRACHEA AND LARYNX	79
1. Preliminary Remarks	79
2. Protocols	80
3. Physiological Principles involved	85
4. Differential Diagnosis between Lodgement in the Trachea and in the Larynx	86
5. Preliminary Preparation for Extraction of the Foreign Body	87
6. On the Technique of the Operative Procedure	88
XI.—LARYNGOTOMY	93
1. Preliminary Remarks	93
2. Principles involved in the Technique	93
3. Treatment of Reflex Phenomena	94
XII.—TRACHEOTOMY	94
1. Preliminary Remarks	94
2. Experimental Evidence	94
3. Practical Application	95
XIII.—INTUBATION	96
1. Preliminary Remarks	96
2. Protocols	96
3. Summary of Experimental Evidence	100
4. Clinical Observations	100
a. Effect upon Respiration	100
b. Effect upon the Heart	100
5. Collapse and Death due to Inhibition	101
6. Differential Diagnosis between Obstruction from Membranes pushed down and Collapse from Reflex Inhibition	101
7. Prevention of Collapse from Reflex Inhibition	103
8. Treatment of Collapse from Reflex Inhibition	103
XIV.—ON THE CAUSE OF CERTAIN PHENOMENA ATTENDING CONSIDERABLE TRACTION ON THE TONGUE	105
1. Preliminary Remarks	105
2. Protocols	106
3. Summary of Experimental Evidence	111
4. Practical Application	113

EXPERIMENTAL RESEARCH

INTO THE

SURGERY OF THE RESPIRATORY SYSTEM

INTRODUCTION

From clinical observations it is apparent that there are a number of phenomena attending operations and injuries of the thorax and the respiratory tract which are not sufficiently well understood for ready application in surgery. Instead of taking up the subject in a general way, it was thought best to divide it into parts, and to make a research on each subject separately. From the nature of the subjects, some parts overlap others. It has been only within recent years that opportunities have been general for experimental work on that part of physiology which relates so directly to surgery. It has been the aim in this work to dwell upon the subjects of most practical importance, and to elucidate, as far as possible, the practical bearings of the several questions under consideration. It is not intended to be exhaustive on all the subjects taken up, nor have all the data accumulated in the experiments been recorded in the protocols, as this would involve unnecessary detail. It was originally intended to make the research both clinical and experimental, but the experimental side grew so large as to make it seem advisable to keep the clinical in the background. It was also intended to present a bibliography of the literature on the various subjects, and the greater part of this has been collected, but will not, at the present time, be added. The experiments have been performed with great care, with the sole desire to arrive at the truth, without reference to previous notions or theories. The research has extended over two years, and was carried out in the physiological laboratory of the Cleveland College of Physicians and Surgeons. In all the experiments dogs were used as subjects, and taken unselected as they were supplied by the laboratory servant. I cannot sufficiently acknowledge my indebtedness to my associate, Dr. W. E. Lower, who rendered most valuable assistance throughout the research, and whose name deserves to appear on the title-page.

METHODS OF RESEARCH AND NOTATION

The animals were all reduced to full surgical anæsthesia before the experiments were begun, and were killed before recovery from the same. In the greater number of experiments ether was employed, and anæsthesia was produced by the following method: A hood was constructed so as to accommodate the animal's entire head; it was made of strong cloth, conical in shape, and into its apex was thrust a piece of cotton-wool. Saturating this piece of cotton-wool with the anæsthetic, and holding the hood closely over the head of the animal, reduction to surgical anæsthesia was made with but little difficulty. After the completion of anæsthesia the trachea was exposed and a breathing canula inserted. To the free end of this canula a strong rubber tubing was attached, and to the end of this tubing was fastened a funnel, which was placed over a piece of cotton-wool saturated with the anæsthetic. By this method anæsthesia

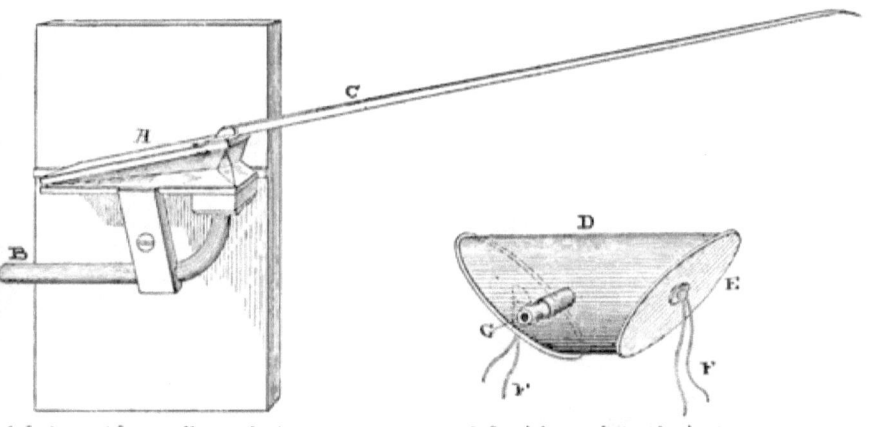

1. Instrument for recording respiration.   2. Receiving respiratory tambour.

was easily maintained and no impediment offered to free respiration, thereby making it possible to carry on the

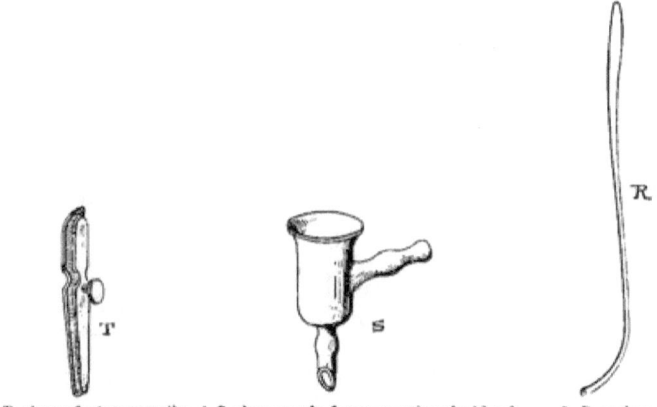

3. T, clamp for temporarily closing vessels.   4. S, glass canula for connecting the blood-vessels with the manometer.   5. R, seeker.

experiments with a minimum of respiratory error. The respirations were recorded by means of a tambour with rubber dam stretched over the ends, in the centre of which was fastened a disk, to which strings were attached; the free ends of these strings were clamped to

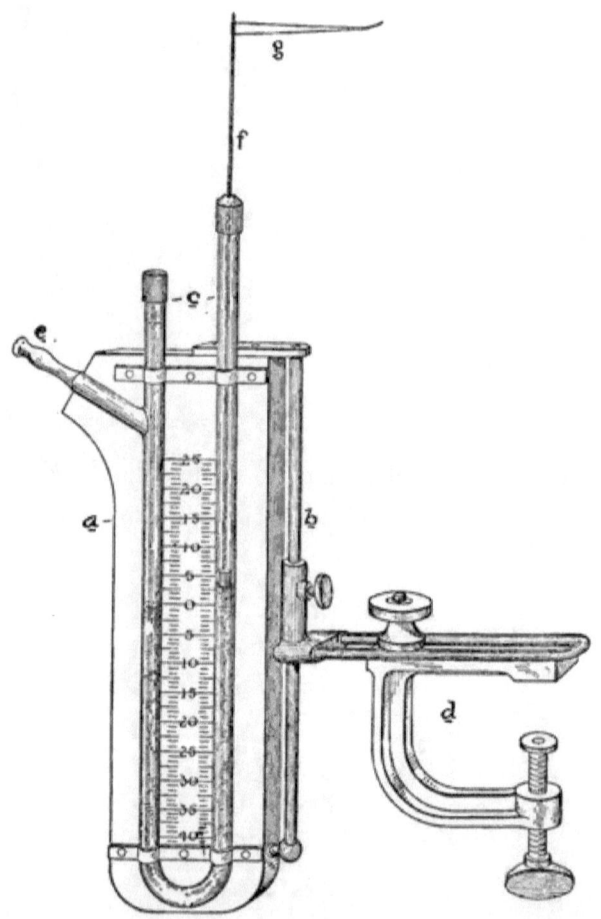

6. Mercurial manometer with graduated scale.

the ends of a piece of duck-cloth, encircling three-fourths of the circumference of the animal's chest. A tube was inserted into the centre of the cylindrical tambour, into the end of which a rubber tube was attached. The free end of this tube was connected with and set in operation by a simple respiratory recording apparatus constructed

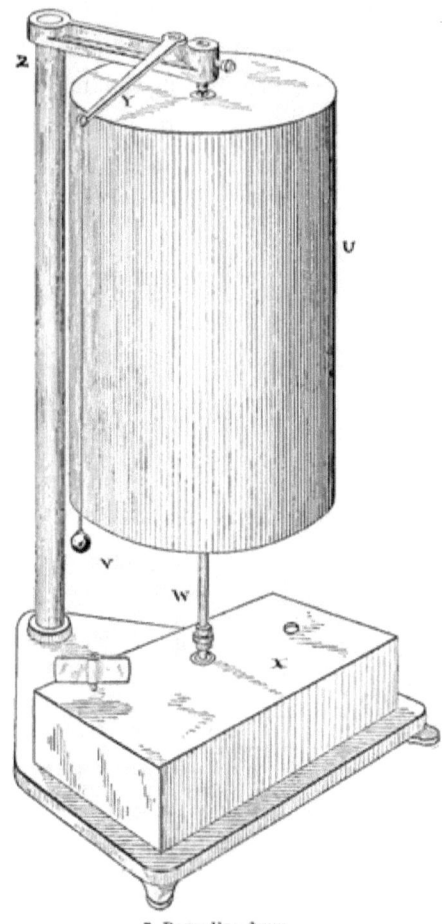

7. Recording drum.

of the material and upon the plan of an organ key. On the upper plate of the organ key was attached a long writing style, armed at its free extremity with a piece of Swedish steel, as thin as paper and tapering to a delicate point, which accurately described upon the smoked drum every phase of the respiratory movement. This method records only comparative respiratory actions. The blood-pressure was recorded in the usual way, by means of mercury manometers upon a revolving drum carrying smoked paper, according to the methods in vogue in experimental physiology. The drums were revolved by a mechanism so made as to be capable of a variety of

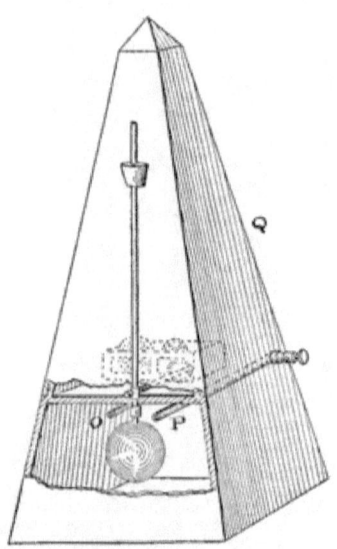

8. Metronome arranged to make and break the electric circuit at O and P in energizing the time-marker.

movements, ranging from one revolution in thirty minutes to eighteen revolutions per minute, so that every phase of any given tracing might be duly recorded. In every experiment tracings were taken, and these have been carefully preserved. Accordingly there is not a statement made in the following pages that may not be verified by tracings in my possession. A sufficient number of typical ones have been published to illustrate groups of their kind. The method of procedure consisted in performing the various operations or imitating the various accidents and conditions while graphic records were being traced, and on each given subject a number of records sufficient to establish or corroborate a law were taken. The apparatus, at least a considerable part of it, was especially constructed to meet the requirements of this particular research. The number of animals subjected to experiments was one hundred and two, and complete notes of all the experiments have been preserved. The notes were made always at the time of the experiment, and every detail was carefully recorded, and from these notes and the tracings the material for this work was obtained. No matter how faithfully the notes and the illustrations may have been made, it is impossible to impart the full impression made by the experiment upon the experimenter, who receives the impressions at first hand.

EXPERIMENTAL RESEARCH INTO THE CAUSE OF THE PHENOMENA ATTENDING THE INHALING OF HOT AIR AND FLAME

Preliminary Remarks.—The sudden development of the phenomena in cases of collapse caused by the inhalation of hot air and flame suggests an essential interference with the physiological function of one or more of the nerve mechanisms of the circulatory or respiratory apparatus. The dogs for the experiments were placed under chloroform anæsthesia. A large blow-flame burner, used in the chemical laboratory for glass blowing, supplied a flame that could be adjusted to a very great range of intensity, and could supply to the animal either flame or hot air.

Protocols.—1. On a bull-dog weighing thirty pounds, with an initial blood-pressure of one hundred and sixty-two millimetres under chloroform anæsthesia, the following experiments were performed: The trachea was dissected up and freed from its attachments, brought forward, severed, and the flame of the blow-pipe made to play directly into its lumen. The immediate effect upon the blood-pressure was a very sharp rise, reaching two hundred and six millimetres. The blood-pressure curve was extremely irregular, recording marked variations with each respiratory movement; the strokes of the manometer were short and rapid, characteristic of an accelerating action on the heart. The respirations at first were shallow and slow, later on became stronger and more rapid, then gradually slowing with great amplitude, finally became smaller and ceased. The blood-pressure remained high during the first ten minutes, after which it rather rapidly fell to the base line. During the time of inhalation the temperature of the animal rose four degrees.

Autopsy.—Examination of the lungs did not reveal any other abnormality than scattered points of slight hemorrhages. The disposition of the blood in the circulatory apparatus did not present any unusual features. The intestines were pale, the kidneys and the liver normal.

2. On a Newfoundland dog weighing forty pounds, with initial blood-pressure at one hundred and fifty-six millimetres under chloroform anæsthesia, the same experiments were repeated. The results were similar, with

the exception that after breathing the flame for three minutes the heart executed long sweeping strokes until death, and the animal continuously breathed the flame for twelve minutes. The respirations at first were shallow, later they became deep, and the inspiratory phase became particularly lengthened and strong.

Autopsy.—Intestines pale, lungs slightly congested; nothing abnormal in the disposition of the blood in the vessels.

3. On a bull-dog weighing twenty-eight pounds, with blood-pressure at one hundred and forty millimetres under chloroform anæsthesia, the same experiments were performed, with results practically identical. In this case the heat was so intense as to melt the adipose tissue around the trachea.

4. On a mongrel cur weighing nineteen and one-half pounds, with the blood-pressure at one hundred and eighty millimetres under chloroform anæsthesia, a hot-air blast was blown through the *mouth* into the respiratory tract, which in this case was allowed to remain intact. Immediately upon contact of the hot air with the pharynx and larynx there were marked "vagal" beats of the heart, with a considerable fall in blood-pressure and a sudden arrest of respiration. The respiratory arrest was but temporary.

Autopsy.—The lungs were slightly congested; the heart, in diastole; there were no clots; the stomach was full of gas; the intestines pale and contracted; the veins dilated.

5. On a healthy poodle weighing ten pounds, with blood-pressure at one hundred and twenty millimetres under chloroform anæsthesia, the following experiments

were performed: The mouth was held wide open and the blow-flame directed into the pharynx and the respiratory tract; the immediate effect upon the blood-pressure was a temporary rise, the heart-beats became slower and fuller, then both vagi were severed, and again the flame was applied as before; the heart now beat very rapidly, and the blood-pressure rose to two hundred and four millimetres, continuing at this high level for some time. The heart continued to beat rapidly and forcibly, and the curve of the blood-pressure was altered by each respiratory act; the rapidity of the heart's action increased until it became so rapid, and the length of the strokes so shallow, that it failed to register on the drum; the respirations at first were much slowed, and for a long time were weak, but afterwards became strong, and remained so to the end of the experiment.

6. On a mongrel weighing twenty pounds, with blood-pressure at one hundred and forty millimetres under chloroform anæsthesia, and with a preliminary injection of one-one-hundredth of a grain of atropine, the same experiments were performed as in the preceding case. The results were the same so far as respirations were concerned, but the blood-pressure rose and no "vagal" beats appeared.

7. In this experiment the preceding manipulation was repeated, with practically the same results.

8. Both vagi were cut and the same experiment repeated as described above; the results were practically the same as in Experiments 6 and 7.

Summary of Experimental Evidence.—The experimental evidence at hand tends to show that the direct effects of flame inhaled into the lungs, with the upper air-passages *excluded*, are not capable of causing sudden death. When the flame is inhaled through the mouth and upper air-passages as well, a very marked reflex inhibitory action upon both the heart and the respiration is produced. In such a case, after a control has been taken (a control showing a characteristic inhibitory action upon the heart and respiration), when both vagi are severed, the inhibitory action upon the heart is wholly prevented, and, while the respirations are not so greatly altered in their character, they do not wholly escape reflex influence. The same may be said of similar experiments in which physiological doses of atropine had been given. The great irregularity of the blood-pressure curve shows that two factors are at work in its production,—the one an accelerating, the other an inhibiting; and in the cases in which the blast of hot air or flame was introduced through the mouth, both of these factors were brought simultaneously into play, while in the experiments in which the flame was forced into the lungs through the trachea, but one factor—the accelerating—was active. In view of the fact that hot air and flame very markedly stimulate reflex inhibition of the heart and the respiration, it is quite probable that the sudden collapse from inhaling hot air or flame is due to the reflex inhibition of the cardiac and respiratory action, in the way just pointed out. Death may be caused in a few minutes by exhaustion of both the respiratory and the circulatory mechanisms from over-stimulation.

RESEARCH INTO THE EFFECT OF FILLING THE CHEST WITH FLUID

Preliminary Remarks.—It occasionally happens that the chest is rather suddenly filled with fluid, and in order to determine the character of the alterations thus produced in the respiratory and cardiac action, animals were subjected to a rapid filling of the chest with normal saline solution at about the temperature of the body.

Protocols.—1. On a bull-dog weighing forty-two pounds, with initial blood-pressure at one hundred and thirty millimetres, the following experiments were performed: A small incision was made in the integument over the anterior aspect of the chest, between the fifth and sixth ribs; then, carefully guarding against the entrance of air, a canula connected with a siphon bottle of salt solution was thrust into the pleural cavity. The solution was allowed to flow rapidly until two thousand five hundred cubic centimetres entered, and the following results were noted: As the chest filled, the blood-pressure at first slightly rose, the character of the heart-beats changed to slow, full beats of extraordinary height, increasing in fulness for about two minutes, then gradually declining until death ensued. The respirations were temporarily arrested, then executed slow and shallow excursions until the end, which was nine minutes after the beginning of the flow.

2. On a water-spaniel weighing twenty-four pounds, with blood-pressure at one hundred and thirty millimetres under chloroform anæsthesia, the following experiment was performed: The technique as described in the preceding case having been carried out, a double hydro-

thorax was induced, producing thereby, as in the former experiment, a gradual decline in the blood-pressure, with

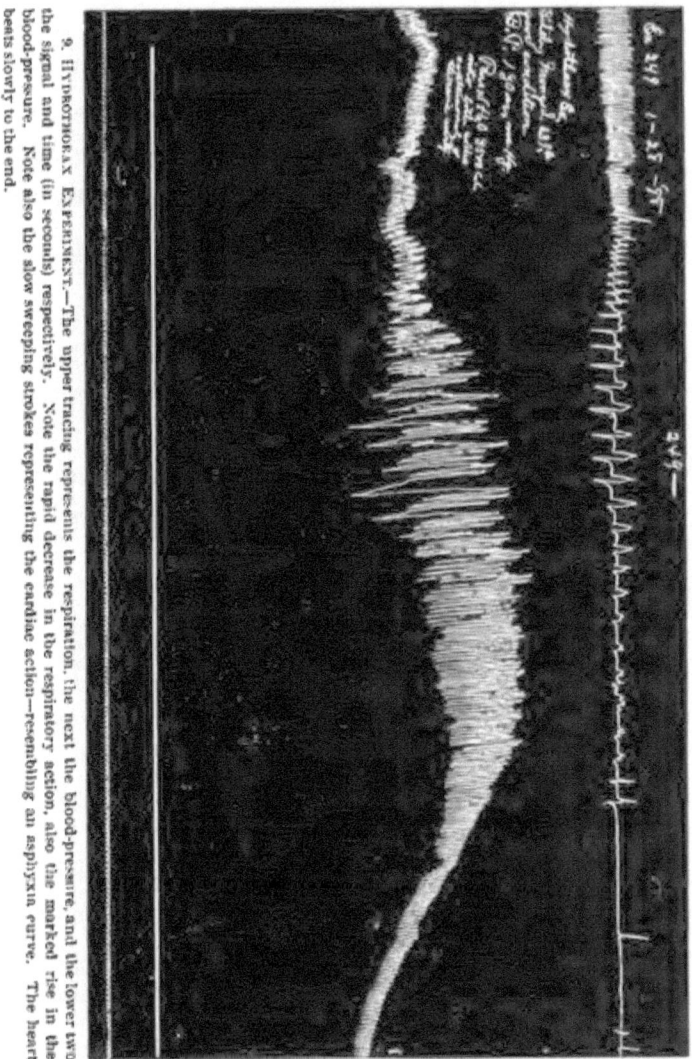

9. HYDROTHORAX EXPERIMENT.—The upper tracing represents the respiration, the next the blood-pressure, and the lower two the signal and time (in seconds) respectively. Note the rapid decrease in the respiratory action, also the marked rise in the blood-pressure. Note also the slow sweeping strokes representing the cardiac action—resembling an asphyxia curve. The heart beats slowly to the end.

slow, full beats, showing very marked inhibition. The respirations were altered in almost precisely the same manner as described in the preceding case. The animal died in twelve minutes.

3. On a shepherd-dog weighing forty-six pounds, with initial blood-pressure at one hundred and forty-eight millimetres under chloroform anæsthesia, the preceding experiments were performed with almost identical results.

4 and 5. Two animals were subjected to experiment under the same conditions as in the preceding cases, with practically the same results.

Summary of Experimental Evidence.—The experimental evidence at hand tends to show that a double hydrothorax, rapidly induced, is extremely fatal. It is impossible for the lungs to become inflated when an additional pressure to that of the natural atmospheric pressure is added to the resistance to be overcome. The phenomena are not the same as in pure asphyxia. In pure asphyxia the blood-pressure suffers a momentary decline, after which there is an actual rise, oftentimes a very considerable rise above its previous height. The character of the heart's action in asphyxia experiments is, during the first several minutes at least, but little altered, thus making a striking contrast to the curve executed by the heart in the experiment upon artificial hydrothorax. While in asphyxia there is for a time powerful respiratory effort, in these experiments a decreasing respiratory action is at once inaugurated. In any given case of sudden hemorrhage filling the chest, the depression produced is only in part due to the pure effects of abstraction of blood from the general circulation. A considerable factor in the result is due to the mechanical effect of fluid interfering with the cardiac and the respiratory action.

RESEARCH INTO THE EFFECT OF PROLONGED MANIPU-
LATION OF THE BRACHIAL PLEXUS AND THE NERVES
SUPPLYING SOME OF THE MUSCLES OF RESPIRATION

Preliminary Remarks.—On several occasions, having observed a dangerous depth of chloroform narcosis rapidly induced in operations involving manipulation of the brachial plexus, a question arose as to whether or not some cause for such a result might be found; a research into this subject was therefore made, to determine whether or not any peculiar effects upon the respiration or the circulation attended operations in this region.

Protocols.—1. On a twenty-pound mongrel, in only fair physical condition, under full *ether* anæsthesia, with the carotid pressure and the respirations registering, after a control tracing had been secured, the following experiments were performed: The brachial plexus was exposed high up, in close proximity to the chest-wall, and continuous manipulating made, causing thereby an irregular circulatory curve, which, on the average, amounted to a slight rise in blood-pressure. At the same time the respirations were increased in frequency and in depth. The increase in frequency was such that five respirations were executed in the time previously occupied by three. With the increased amplitude and increased frequency there was a very marked increase in the total volume of air exchanged in the lungs. This manipulation was continued for ten minutes, and the respiratory increase was sustained. On cessation of the manipulation the respiratory rhythm was reduced in frequency and amplitude in about the same ratio that the manipulation had caused the increase. On exposing the opposite plexus and repeating like procedures, equally striking results were obtained. It was observed, however, that when the principal traction was made upon the peripheral nerve-trunks, the effect upon the respiration was very much less than when the traction was made on the central portion of the nerve-trunks. Then, allowing the animal a little time for rest, an assistant repeated the manipulation on the one side, while on the opposite side similar manipulation was made by myself. It was found that the more the central portion of the trunks was stretched and manipulated the more marked was the respiratory excitation, and that if during this manipulation the anæsthetic was inhaled in its usual quantity, an excessive anæsthesia was induced. After a considerable lapse of time the manipulation and traction produced less marked respiratory alterations. The nerves supplying the muscles of the chest were then likewise subjected to manipulation, producing an increase in the frequency and amplitude of the respirations, though not so striking as the preceding.

2. Under *chloroform* anæsthesia, but in other respects under similar conditions to the preceding case, a fifteen-pound mongrel was subjected to similar experiment. After securing a control, the nerves were again subjected to continued manipulation, causing a marked increase of the respiratory action and very irregular blood-pressure curve. Very soon, however, the inhalation of the usual amount of chloroform produced an over-anæsthesia, and the blood-pressure rapidly sank. On cessation of the manipulations the respirations diminished very markedly in frequency and amplitude. The animal was reduced

almost to collapse, from which it required considerable time, together with artificial respirations, to restore it. After some time had elapsed and restoration was completed, the same procedures were repeated with like results.

3. On an eighteen-pound bird-dog the following experiments were performed under conditions similar to the first: The skin was rapidly removed from the chest, exposing the pectoral muscles. Manipulation was then begun upon the muscular structures and the nerves of extraordinary muscles of respiration. This manipulation produced an increased respiratory action, which increased in frequency and amplitude as before. The blood-pressure, however, was much less altered than in the preceding case. On cessation of this manipulation, the respirations again resumed the normal rate. On allowing the chloroform cone to remain in position, then manipulating the nerves and muscles on both sides of the chest simultaneously, an over-anæsthesia was very rapidly induced. The over-anæsthesia was first observed in the marked depression of the blood-pressure. After the cessation of the manipulations the respiration became very slow and shallow, showing likewise the effect of the over-anæsthesia.

4. A mongrel, weighing nineteen and one-half pounds, in good condition, whose carotid blood-pressure was one hundred and eight millimetres, was subjected to a simultaneous dissection, irritation, and traction on both brachial plexuses and nerve-muscle apparatuses. These mechanical stimulations produced a very marked increase in the respiratory action, more than doubling that recorded in

the control. Anæsthesia was carefully guarded during the first portion of the experiment, but in the latter portion the anæsthetic was allowed to remain with the degree of chloroform saturation that was believed to be safe in sustaining an ordinary anæsthesia. On repeating bilateral simultaneous manipulations, the respirations were again very markedly increased in both frequency and the amplitude, thereby inducing very rapidly an over-anæsthesia. This over-anæsthesia was first observed in the rapid decline of the blood-pressure, and later in the respiration. When this effect was noted, on cessation of the manipulation there was a rather rapid failure of the respiratory action. Artificial respiration was necessary to restore the animal. After this restoration the dog was so weak that it did not readily respond to further experiments.

5. In this experiment the foregoing technique was repeated. There was, however, but little alteration in the respiratory action, even though the manipulation was quite as severe as in the preceding case.

Summary of Experimental Evidence.—It was found that during manipulation of the brachial plexus, or of the nerves supplying the muscles of respiration, an increased respiratory action was produced, resulting in over-anæsthesia by causing an excessive inhalation. This was specially true in experiments in which chloroform was administered. On the cessation of the manipulation the normal respiratory action was resumed. It was found, however, that by continuing the manipulation of these structures there was a greater tendency towards respiratory failure later on in the experiment. It is true that mechanical stimulation of any nerve will produce increased respiratory action. It would seem that the manipulation of the respiratory nerve-muscle apparatus or nerve trunks may affect the respiratory action to an extent sufficient to constitute a danger of over-anæsthesia, provided the inhaled air is saturated with the anæsthetic to the same degree as in the control. In such experiments an earlier break-down of the respiratory mechanism occurs.

On cessation of a severe manipulation, respiratory depression or failure is very likely to occur, because then, the mechanical stimulation having been withdrawn, the mechanism is thrown upon its own resources, which have been much impaired by the previous stimulation.

Practical Application.—The depression referred to in the preliminary remarks would probably appear only in cases involving rather direct manipulation of these nerve-structures, and more readily in chloroform anæsthesia on account of the relatively greater potency of this anæsthetic.

But the point of greatest practical importance is the proper interpretation on the part of the anæsthetizer of the cause of the increased action. Such increased action would ordinarily be interpreted as a symptom of under-anæsthesia, and at once suggest giving more anæsthetic. Under these circumstances the anæsthetic effect would be more marked upon the circulation than upon the respiration until the patient sinks into a collapse more or less profound. The respiratory depression appears later because of the mechanical stimulation which will maintain respiratory activity until the centres are exhausted.

RESEARCH INTO THE CAUSE OF COLLAPSE OR DEATH FROM BLOWS UPON THE LOWER CHEST AND THE EPIGASTRIUM

Preliminary Remarks.—This research was suggested by the number of cases of collapse and even death following blows delivered upon the chest, more especially over the lower portion of the left anterior chest-wall and upon the so-called "pit of the stomach."

In the history of these cases it is usually stated that such results instantly followed the application of the violence, and that if death did not immediately follow, recovery took place.

The so-called "short rib" and "solar plexus" blows, well known in pugilistic encounters, are examples.

The following have been the principal theories advanced as to the manner of the production of the effect: That the blow produced mechanical stimulation of the solar plexus, thereby causing cardiac arrest; that the diaphragm was injured, causing respiratory arrest or spasm of this organ; that the heart itself was injured; and that the heart-failure was due to mechanical stimulation of the vagus.

Protocols.—1. A shepherd-dog, weighing twenty pounds, in good physical condition, was placed under full ether anæsthesia, and arrangements made for securing graphic records of the respiration and of the carotid blood-pressure. After control tracings had been secured, a severe blow was delivered with a rather heavy handle of a hammer directly upon the chest over the point of appearance of the apex-beat. The heart was instantly arrested and the blood-pressure fell to the abscissa line, from which it did not again rise. The heart did not make a single observable effort towards resuming its function. The respirations were temporarily completely arrested, then there were several rather feeble respiratory efforts, becoming more shallow and appearing at longer intervals until they entirely failed. The animal instantly died from the effect of the single blow delivered in the manner above described.

Autopsy.—The heart was found in diastole, all the chambers containing more or less blood. There was no observable lesion of the pericardium or of the heart-muscle.

2. On a water-spaniel, weighing eighteen pounds, in fair condition, under surgical ether anæsthesia, with arrangements for graphic records, as in the preceding, a direct blow was delivered upon the chest over the heart. The blood-pressure underwent an immediate staggering fall; the heart-strokes were slow, full, but irregular; the blood-pressure curve was correspondingly irregular, but at the end of sixty-five seconds it had gained the same height it occupied before the blow was delivered. The respirations were momentarily arrested, and they became irregular, but their amplitude was almost doubled. This increased amplitude and increased pause with irregularity continued during the space of time in which eight respiratory excursions would have been made, on the basis of the same rate at which the respiratory actions were performed at the time of delivering the blow. The same experiment was subsequently twice repeated, and in each case with practically the same results. The left vagus was then sev-

ered, after which there was an increase in the respiratory depth and no effect observed upon the circulation. On severing the right vagus, respirations ceased for a time equal to that required for three normal respiratory excursions, then the respirations were gradually increased in their amplitude, their rapidity remaining about the

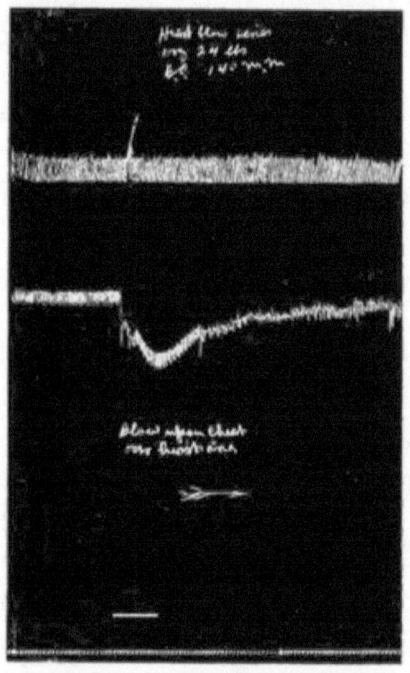

10. BLOW UPON THE CHEST OVER THE HEART.—Note the abrupt fall in the blood-pressure (second tracing), and the irregular heart-action afterwards.

same. The heart was arrested for three seconds; then, after a few slow beats, it increased very considerably in the rapidity, and the excursions of the manometer were diminished in length. On repeating the direct blow upon the chest quite as forcibly as before, an interference with the respiration and the circulation, similar to that noted previous to section of the vagi, was observed. The blows were twice more delivered, each time producing like respiratory and circulatory phenomena.

3. A water-spaniel, weighing twenty-five pounds, under ether anæsthesia, with arrangements for records as in the preceding case, was subjected to the following experiments:

(*a*) After securing control tracings, a blow from a light hammer was directed upon the chest over the heart, causing thereby a sudden drop in the blood-pressure, with temporary cessation of cardiac action, followed by sweeping beats at about twice the interval observed before the experiment. The blood-pressure curve during this time was irregular, and in thirty-six seconds had risen to the level it occupied before the experiment. The respirations were irregular during eight seconds.

(*b*) The same experiment was repeated with like results.

(*c*) Both vagi were severed, producing variations during the performance of this operation similar to those described in the first experiment, so that after both had been severed the heart beat more rapidly, the blood-pressure slightly rose, the cardiac strokes were shorter, the respirations deeper and a little slower.

(*d*) A blow was now delivered upon the chest over the heart, as nearly like preceding as possible. There was practically no effect on the respirations, but the blood-pressure failed suddenly as before, the character of the beats was the same, and the blood-pressure was regained in about the same time.

(*e*) The same experiment was repeated with practically the same results.

4. On a shepherd-dog weighing twenty pounds, under ether anæsthesia, with arrangements for records as in the preceding, the following experiments were performed:

(*a*) A blow upon the chest directly over the heart produced a marked fall in the blood-pressure, with "vagal" beats continuing for some time; then the "vagal" beats suddenly gave way to small, shallow, and extremely rapid beats. The respirations were temporarily irregular, but not greatly disturbed.

(*b*) Both vagi severed, producing the phenomena described in Protocol 1, and when both had been severed the blood-pressure rose, while the respirations became deeper and slower.

(*c*) A blow delivered upon the chest over the heart, as in (*a*), did not produce any alterations in respiration. There was a sudden fall in the blood-pressure; during this time the heart executed three slow "vagal" beats, then quickly regained the normal level of pressure.

(*d*) Injected one-one-hundredth of a grain of atropine into the jugular vein; after the lapse of one minute, stimulated the peripheral end of the severed vagus by applying the electrode of a Du Bois-Reymond coil, whose secondary coil was overlapping the primary by one-half. This electrical stimulation, though more than sufficient to cause inhibition of the heart, did not affect a single beat. This was done as a control to test the full physiological action upon the terminals of the vagus in the heart, so that if, in a subsequent experiment, the heart again was affected, it could not be due to a mechanical stimulation of the vagus, either its branches or its terminals.

(e) A blow as nearly similar as possible to those previously given, was then directed upon the chest over the heart, producing a sudden fall in the blood-pressure with four beats of like character to those described under (c),

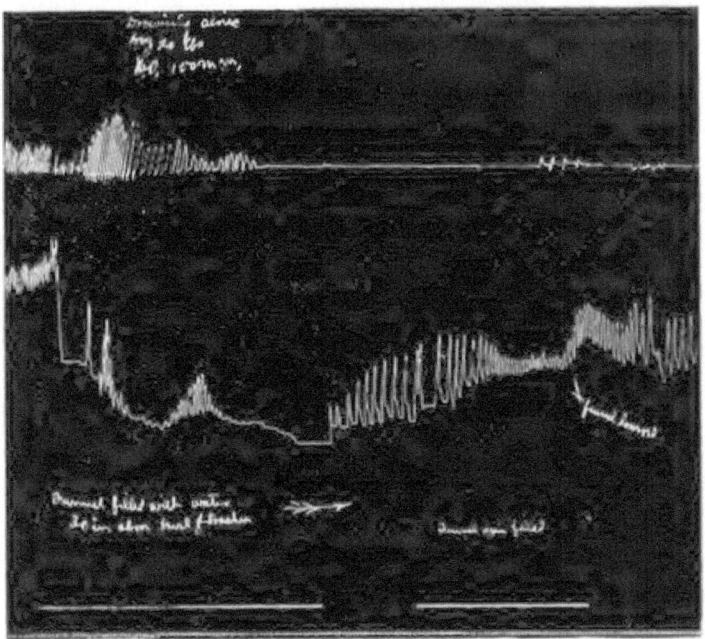

11. The upper tracing represents the respiratory action, the next the blood-pressure. At the lower margin are the signals and time-marker (seconds). Note the momentary rise, followed by the collapse in the circulation after the lungs were subjected to pressure of thirty inches of water.

the blood-pressure regaining its previous level after executing a curve in all respects similar to that in (a). The respirations were not affected.

(f) The same experiment repeated with like results.

5. On a healthy mongrel dog weighing eighteen pounds, under conditions as nearly like the preceding as possible, like experiments were performed with results essentially the same,—that is to say, after control ex-

periments the vagi were cut, and after testing the effects with severed vagi, atropine was given as in the preceding; then the same experiment was repeated, obtaining practically the same results.

6. On a mongrel dog weighing fifteen pounds, under conditions similar to the preceding, the following experiments were performed:

(*a*) A control blow over the right side of the chest was delivered, causing thereby temporary irregularity in the respiratory action and a marked fall in blood-pressure, produced by a single intermission.

(*b*) Similar effects were produced by delivering blows with considerable force over various portions of the chest. However, the nearer the heart area the more marked was the effect.

(*c*) With the vagi cut similar blows produced like results upon the circulation, but no effect on respiration.

7. On a shepherd-dog weighing twenty-five pounds, under conditions similar to the preceding, the following experiments were performed:

(*a*) A blow delivered with great force over the "pit of the stomach" produced a sudden drop in the blood-pressure, continuing during the time of a single beat. The respirations were but momentarily altered.

(*b*) A similar blow upon the abdomen over the umbilicus produced no effect.

(*c*) A blow was delivered over the kidneys and the lower abdomen, but produced no effects.

(*d*) As a control, a blow was again delivered upon the chest over the heart, producing a more gradual though marked fall, which was recovered in ten seconds.

8. On a thirteen-pound mongrel dog, under conditions like the preceding, the following experiments were performed:

(*a*) As a control, a blow was delivered over the heart with the usual results.

(*b*) A severe blow was delivered over the "pit of the stomach," with results similar though less pronounced.

(*c*) The stomach was freely exposed by a median incision and a considerable quantity of air forced into it through the œsophagus, then the œsophagus was securely tied. The pylorus was also securely tied. A blow was then delivered directly upon the stomach, with its force directed a little upward, producing a fall in blood-pressure similar to, though less marked than in (*a*).

(*d*) The stomach blow was repeated after section of the vagi with similar results. Also after the injection of one-one-hundredth of a grain of atropine into the jugular vein, with similar results.

9. On a yellow mongrel dog weighing fifteen pounds, under conditions similar to the preceding, the following experiments were performed:

(*a*) Placing the dog upon his right side, a severe blow was delivered upon the chest in the left axillary line, producing results similar to the blow directly over the heart, though less marked.

(*b*) Placing the dog upon the left side, a similar experiment was performed upon the right, with similar results, though less marked in degree. A blow delivered directly over the heart caused but a momentary drop in the blood-pressure, but this drop was deeper than in the case in which the blow was delivered upon the side. It was found on repeating the blows, even with increasing severity, that the effect upon the blood-pressure was in each subsequent experiment lessened. At the autopsy it was found that the heart itself was partially ruptured and the pericardium contained a considerable amount of blood; three ribs had also been broken by the violence of the blows.

10. On a sixteen-pound yellow mongrel, after making a free exposure of the diaphragm by an abdominal incision and removing the stomach from its relation to this organ, direct pressure was exerted upward against the apex of the heart, producing thereby a very marked arythmia and a very irregular blood-pressure curve, which, on the average, amounted to a fall. These phenomena continued as long as the mechanical interference through the diaphragm continued. It was found that even comparatively slight pressure upward upon the diaphragm against the apex of the heart produced great disturbance in the cardiac action. This disturbance was characterized by a very great irregularity in the force of the beat, in the length of the strokes, and in their rhythm, producing thereby a very irregular blood-pressure curve. Placing the end of a hammer over the diaphragm at the point where the apex impulse could be made out, then delivering a blow upon the distal end of the hammer, produced a staggering fall in the blood-pressure, with very marked irregularity of the heart-beats. Repeating this same manœuvre on the right side of the diaphragm, away from contact with the heart, no cardiac effect was noted. Likewise repeating the same on the extreme left

border of the diaphragm, directing its force downward towards the ribs and therefore away from the heart, no cardiac disturbance was noted.

11. A young water-spaniel, weighing sixteen pounds, was subjected to experiments similar to the preceding, producing results in all respects similar. In addition to these, the closed hand was placed against the abdominal side of the diaphragm, gradually pushing this organ upward; at first no effects were produced, but when this pressure was increased and the area of cardiac activity was further encroached upon, irregularities in action similar to the preceding were observed. The chest-wall was then cut away over the heart and a smart blow was delivered directly upon that organ, producing a fall in blood-pressure and an alteration in the cardiac action, similar to those heretofore described.

12. A savage bull-dog, weighing forty pounds, was subjected to the following experiments: Artificial respirations were maintained, and the thorax over the heart was cut away so as to expose it. Blows were then delivered upon the heart itself, producing effects entirely similar to those noted in the preceding experiments. The pericardium was then incised and the heart-muscle itself subjected to similar treatment, producing similar results. Any interference with the heart, as an organ, produced a staggering fall in the blood-pressure, and, on the withdrawal of the mechanical stimulation or interference, the blood-pressure curve would rapidly rise again. This very soon exhausted the animal.

13. A shepherd dog in good condition, weighing fourteen and one-half pounds, was subjected to the same pro-

cedure as in the preceding case, with, in the main, the same results.

14. A mongrel-dog, weighing fifteen pounds, in good condition. After a control had been taken, an incision was made through the abdomen, and the viscera were withdrawn so as to expose the solar plexus. A blow was delivered upon the plexus, the force of which was so direct as to interfere as little as possible with the large blood-vessels. No effect upon the heart was noted. The solar plexus was then grasped between the fingers and manipulated as rudely as possible, without producing any direct effect upon the heart. During this manipulation, however, there was a gradual decline in the blood-pressure *pari passu* with the dilatation of blood-vessels in the splanchnic area. Finally, as a control, a blow was delivered over the heart, producing the changes in cardiac action and blood-pressure described in previous experiments.

15. On a sixteen-pound bird-dog the same experiments were performed as in the preceding case, with practically the same results. No amount of manipulation or mechanical injury of the solar plexus or the splanchnic trunks produced any sudden change in the blood-pressure or in the cardiac action. It did produce a gradual decline in the blood-pressure, as in the preceding experiment. The control blow delivered directly upon the heart caused the usual results, though to a rather light degree.

Summary of Experimental Evidence.—No amount of injury inflicted upon the solar plexus, either directly or indirectly, was capable of causing any inhibitory action upon the heart, and in no way did such injury contribute to the immediate death or collapse referred to, but the effect this manipulation did have was to produce a vaso-dilatation of the "splanchnic area," thereby causing a gradual decline in the blood-pressure.

In experiments in which the abdomen was opened and the diaphragm protected from indirect violence, blows upon the stomach alone produced but little effect upon the blood-pressure or the respiration.

In other experiments in which the diaphragm was not protected from indirect violence transmitted through the stomach or other viscera, blows delivered upon the pit of the stomach, especially when the stomach was distended, produced sometimes a momentary staggering fall in the blood-pressure. In some cases the pressure remained at a lower level and described a very irregular curve, then finally regained the height of the control.

Pressure suddenly applied, or blows directly upon the diaphragm within the cardiac zone, produced usually a very marked drop in the blood-pressure. In some cases but a single intermission of the heart-beat was produced, in others the blood-pressure suffered a great fall, and slowly and with great irregularity of action regained the lost pressure. Even carefully pressing with the hand upward against the diaphragm so as to produce by such pressure an interference with the freest movement of the apex caused a very great cardiac irregularity and an irregular blood-pressure.

Blows delivered upon the lower chest, especially over the cardiac area, produced various results. In one case the heart was instantly arrested from the effect of a smart but not heavy blow from a small hammer over the cardiac area, and death at once ensued. Respiratory action was as suddenly arrested. The most common result was a very great drop in the blood-pressure,—a collapse. Then after a variable time the heart would regain its normal action and the lost blood-pressure would be recovered.

The effects, as nearly as observations would permit deductions, varied in different dogs, the blow being delivered with about the same force, with the same instrument, and, as nearly as may be, at the same point of application. On the whole, it may be said that the more nearly the blows were delivered over the centre of the area of the cardiac dulness the greater the effect.

Blows delivered upon the naked heart *in situ* produced like results, though more profound.

The evidence thus far offered tends to show that the solar plexus may be disregarded as a factor, and that the cause of the striking phenomena is the mechanical effect of violence upon either the heart-muscle itself or upon its nerve-mechanism. The effect might have been induced by a "vagal" or inhibitory action. If so, it was either a direct or a reflex inhibition,—interpreting direct inhibition as that due to an excitation of the trunk of the vagus or its cardiac branches, and reflex as afferent impulses sent up to the centre in the medulla from irritation of some branch of the vagus,—*e.g.*, the superior laryngeal.

In the animals in which both vagi had been previously severed, when the foregoing experiments, blows, etc., were repeated, like results were produced. The collapse, then, in these cases could not be due to reflex inhibition, because the path by which the afferent impulses reach the heart had been severed.

In the experiments in which atropine in sufficient dosage to paralyze the terminals of the vagi in the heart was given, and this dosage was proved by applying a faradic current from a Du Bois-Reymond apparatus to the vagi while the secondary coil was lapping the primary by half, upon repeating the foregoing experiments like results were obtained, though, as nearly as could be estimated, the collapse was not so prolonged as before.

Collapse or death may be caused wholly independently of the vagi, though the vagi probably slightly contribute to the result.

Finally, collapse or death from violence applied upon the lower chest or abdomen are due mainly to the loss of rhythmic contractions from the mechanical irritation of such violence on the heart-muscle itself. There is evidence tending to show that the vagal terminal mechanism in and near the heart contributes to the result, but in a minor degree.

The practical deductions are sufficiently obvious, and need not be here repeated.

EXPERIMENTAL RESEARCH INTO THE MECHANISM OF DROWNING

Preliminary Remarks.—It has been observed that in certain cases of drowning death has been almost instantaneous, while in other cases a considerable time has elapsed. In the cases of sudden drowning, although the body may be immediately recovered, no sign of life remains. This research was undertaken to discover an explanation for this discrepancy in the length of time before death occurs, and also, if possible, to throw some light upon those cases of sudden death in good swimmers, who were supposed to be able to take care of themselves in the water.

In the first series the animals were wholly submerged in shallow water and allowed to drown.

In a second group of experiments the trachea was entirely severed, and a large canula tied in its open end; to this canula a rubber tubing was connected, varying in length from two to four feet, into the end of which a funnel was attached. With this apparatus, by pouring water freely into the funnel, an effect could be produced fairly comparable with that of submergence in water to a depth equal to the difference between the level of the lungs and the height of the water in the funnel.

In the third series of experiments the trachea was severed, the animal allowed to breathe through a canula placed in the trachea, and water forced through the mouth and upper air-passages, but not reaching the tracheo-pulmonary tract.

In the fourth series water was thrown with considerable force into the animal's pharynx, with mouth wide open and tongue drawn forward, so as to allow direct contact with the upper air-passages, the trachea remaining intact.

In the fifth series preliminary injections of physiologic doses of atropine were made, so as to estimate the part played by the vagi.

In the sixth series both vagi were severed, preliminary to the drowning. In some of these experiments two animals were simultaneously drowned under different conditions, so that comparisons might be made.

Protocols.—1. On a Skye terrier twenty-one pounds in weight, in rather poor condition, with control blood-pressure of one hundred and seventy-two millimetres, under ether anæsthesia, the following experiments were performed: After having placed the respiratory apparatus upon the chest and introduced the carotid canula to mark the blood-pressure and secure a good control, the entire animal was submerged, to the extent of merely covering the body, in water at 97° F. There was some fall in the blood-pressure at once; then the heart-beats were slowed, executing long strokes, and the blood-pressure rose again to the height it was in the control. The blood-pressure remained at this height during forty-five seconds. Then it gradually fell, the heart-beats became slower and more shallow until death. The heart continued beating nine minutes. The respirations at the onset of this submersion were immediately arrested; but little respiratory action was noted during the first twenty seconds, after which there were powerful respiratory efforts for two and one-half minutes; then the respirations became less powerful and less frequent, gradually decreasing, and failed before the heart did.

Autopsy.—The intestines were very much congested and in slight contraction. The venous trunks were everywhere full. The right auricle was greatly distended, the right ventricle much less so; the left auricle was full, and the left ventricle empty and in contraction. The posterior lobes of the lungs were discolored, and on being thrown into water they floated low.

2. On a rat-terrier weighing fifteen pounds, with control blood-pressure at one hundred and forty millimetres, under ether anæsthesia, the following experiments were performed: The preliminary preparations were made as in the preceding case, excepting the temperature of the water, which was 52° F. The animal lived seven minutes. On submersion there was only a slight immediate fall in blood-pressure, then there followed a rise to a point higher than the control pressure. The character of the heart-strokes at the time of the beginning of the rise was changed, the manometer executing sweeping excursions with a slower rhythm than in the control,—that is to say, there was a distinct "vagal" action. The blood-pressure remained at this elevated level for two minutes and fifty seconds, then it declined until death. In this experiment the "vagal" character of the heart-beats was the prevailing characteristic. The respirations at first were very considerably inhibited, then they grew stronger between the first and third minutes, after which they became irregular, slower, and weaker until death. Respiratory and cardiac action failed almost simultaneously.

Autopsy.—The posterior lobes of the lungs displayed a purplish discoloration. The lungs floated on being thrown into the water, although they floated rather low; the heart was in diastole, right auricle and right ventricle more distended than the left. The left auricle was more distended than the left ventricle. The stomach was extremely distended with air. Venous trunks everywhere full; blood dark.

3. On a mongrel cur, weighing fifteen pounds, with control blood-pressure at one hundred and ten millimetres,

under chloroform anæsthesia, the following experiment was performed, as nearly as possible under the same conditions as in the preceding case. The blood-pressure on submersion of the animal remained at the same level. The "vagal" beats, as above described, appeared in the latter part of the first minute, and were continued until death. In the third minute the heart ceased beating temporarily, but in ten seconds regained its beat as before. The respirations were at first slowed, then temporarily arrested, but soon began a vigorous action, reaching the maximum in the third minute, then gradually failing, the animal dying in seven minutes.

Autopsy.—The immediate autopsy revealed the heart in diastole. The right side was more distended than the left; clots in both sides. The venous trunks were everywhere full; the arterial system almost empty; intestines congested.

4. On a mongrel weighing twenty-four pounds, with control blood-pressure at one hundred and twenty millimetres, under chloroform anæsthesia, the same experiment was performed as in the preceding case. The temperature of the water in this case was 50° F. The animal lived ten minutes. Immediately on submersion there was a slight rise in the blood-pressure, without alterations in the character of the heart-beats. This was soon followed by a marked fall in the blood-pressure and the appearance of "vagal" beats, as above described. The fall, however, was but slight, and during the first five minutes the blood-pressure was well sustained. Later in the experiment the heart showed great irregularity in its action, executing a number of rather rapid strokes, then a num-

ber of extremely slow ones. The respirations were as gradually slowed and became quite shallow, then they became irregular, and later on were followed by strong rapid efforts, these finally diminishing in both force and frequency until complete cessation. Respiration stopped about forty seconds before the heart did.

Autopsy.—Considerable water in the lungs, slight discoloration of the posterior lobes, the heart in diastole, the venous system full, the arterial empty, intestines congested.

5. A shepherd-dog, weighing forty-seven pounds, with control blood-pressure at one hundred and thirty millimetres, under chloroform anæsthesia, with the water at the temperature of 97° F. The same experiment was performed as in the preceding case, with practically the same results, the heart beating during nine minutes.

Autopsy revealed conditions similar to the preceding case.

In the following group of experiments the animals were drowned by introducing water into the pulmonary tract through the trachea.

6. On a shepherd-dog weighing twenty pounds, under ether anæsthesia, the following experiment was performed: The trachea was dissected out and severed just below the cricoid. Into the free end a canula was inserted and closely tied; to the canula rubber tubing was attached, and after control tracings had been secured, the free end of the tube was immersed in water at a temperature of 128° F. There was a direct fall of fifteen millimetres in the blood-pressure after the first inspiratory effort. The respirations were shallow and slowed. The tube was

kept under the water for twenty seconds, and, when removed, the animal rapidly regained the lost depression. The same process was again repeated during twenty seconds, with the same results. The process was four times repeated, and practically the same results followed. It was found that, if the experiments were continued until the blood-pressure became quite low, and if the animal were allowed to breathe air again, the blood-pressure would rapidly recover.

Autopsy.—The chest was at once opened; the venous trunks were all much distended, the right side of the heart full, the left side contracted. The blood clotted very quickly; veins in the splanchnic area distended; all the blood dark.

7. On a spaniel weighing twenty-two pounds, with control blood-pressure at one hundred and forty millimetres, under ether anæsthesia, the following experiments were performed: With the preliminary preparation the same as in the preceding case, the tube was again placed under water and the dog allowed to breathe nothing but water for one minute. There was an immediate fall in blood-pressure, after which there was a compensatory rise, with exhibition of long strokes. On allowing the animal to breathe air, the blood-pressure curve again recovered itself. The respirations at first were temporarily arrested, then became shallow and irregular, but finally grew more powerful and more frequent. This experiment in almost every detail resembled the preceding.

The autopsy showed the venous trunks everywhere distended, the right side of the heart full, the left side empty, blood very dark.

8. On a skye terrier, weighing twenty-two pounds, with control pressure at one hundred and fifty-four millimetres, under ether anæsthesia, the following experiment was performed: To the end of the rubber tubing three feet long a large funnel was attached, and into this funnel water was poured at a temperature of 148° F., so as to keep more or less water in the funnel all the time. There was a momentary sharp rise in the blood-pressure; this was followed by a sharp decline of ten millimetres, then a sudden drop, almost to the abscissa line of the central blood-pressure. After this sudden drop, which was accomplished within five seconds, there was after thirty seconds a compensatory rise in the blood-pressure, but the heart-beats had practically ceased at the end of sixty seconds. The compensatory rise was but small in comparison with the great fall. The heart-beats that appeared after the animal's collapse were slow and feeble, the respirations were instantly temporarily arrested, but after from ten to thirty seconds they were renewed and continued to increase in amplitude for about thirty seconds, when they rather rapidly ceased. The animal was practically dead, so far as may be estimated from a total collapse of the blood-pressure, within five seconds after the water had filled the lungs and the rubber tubing.

Autopsy.—An immediate autopsy showed the venous trunks greatly distended, the right side of the heart full to over-distention, the left side of the heart empty and contracted, the blood very dark.

9. On a mongrel weighing fourteen and one-half pounds, with control blood-pressure at one hundred and forty millimetres, under ether anæsthesia, the following

experiment was performed: The technique of the preceding case was carried out in every detail, with the exception that the water was introduced at a temperature of 36° F. In this case there was a momentary sudden rise in pressure, then a very considerable fall, lasting during five seconds, after which there was a great collapse in the blood-pressure, which sank almost immediately to the abscissa line. The cardiac action after this short time was represented by a few feeble strokes. The circulation after the lapse of these few seconds was completely arrested. As to respiration, there were five slow respiratory efforts, then for seven seconds there was no respiration at all, and afterwards the respiratory efforts were slow, diminished in amplitude, and at the end of fifty-six seconds entirely ceased. During the following two minutes there were occasionally gasping respiratory efforts.

Autopsy.—Immediate autopsy showed all the venous trunks full, arteries quite empty, right side of the heart extremely distended, left side empty and contracted.

10. On a mongrel cur weighing twenty-four pounds, with control pressure at one hundred and twenty-eight millimetres, under ether anæsthesia, arranged as in the preceding case, a similar experiment was performed, holding the funnel at the same height as before. After the water began to flow the heart was slowed during ten beats; during this time there was a considerable fall in the blood-pressure, then there was a collapse during four seconds, approaching nearly to the abscissa line. This striking fall was most noted during the respiratory efforts. During expiration the blood-pressure rose considerably; during inspiration it correspondingly fell. Respirations

were at first slowed and more shallow, afterwards greatly slowed, with strong, continued expiratory efforts. Respirations, however, stopped almost simultaneously with the heart. The animal was practically dead at the end of nine seconds.

Autopsy.—An immediate examination showed the venous trunks distended, the right side of the heart full and flaccid, left side empty and contracted.

11. On a bull-dog weighing forty-four pounds, in good health, with control blood-pressure at one hundred and fifty-five millimetres, under ether anæsthesia, with conditions the same as in the preceding case, the following phenomena were observed. After the water began to flow into the lungs there were five sweeping heart-beats, during which, after a momentary rise, there was a considerable fall in the blood-pressure, and after these five beats had been executed there was a collapse in the blood-pressure. This great collapse occurred during the inspiratory phase of respiration. After a lapse of a short time there was a slight compensatory rise in the pressure; this rise was comparatively small. The respirations at first were slowed, then became more feeble and less frequent until death.

Autopsy.—An immediate examination showed the venous trunks distended, the right side of the heart full and flaccid; the auricle was more distended than the ventricle, the left side empty and contracted.

12. On a water-spaniel weighing twenty-eight pounds, in fair health, with control blood-pressure at one hundred and forty-two millimetres, under ether anæsthesia, the preceding experiment was repeated. The blood-pressure

curve showed all the characteristics displayed by the preceding case, excepting the collapse was not so sudden and the compensatory rise was greater. Death, however, occurred within a few seconds. The respiratory phenomena were practically the same as in the preceding experiment.

At the autopsy, the venous trunks were all distended, the blood was extremely cyanotic, the right side of the heart full and flaccid, especially the right auricle, the left side empty and contracted.

13. On a nineteen-and-one-half-pound mongrel, with control blood-pressure of one hundred and fifty-four millimetres, under ether anæsthesia, under conditions similar to the preceding, and with the same technique, results were obtained in almost every respect similar, excepting that the respirations were more nearly completely arrested at the onset. There was again the same collapse in the blood-pressure and the quick death.

Autopsy.—An immediate examination showed the venous trunks distended, the right side of the heart full and flaccid, left side empty and contracted. The lungs, when thrown into water, floated low.

14. On a forty-seven-and-one-half-pound bull-dog, with control blood-pressure at one hundred and forty-five millimetres, under ether anæsthesia, and according to the technique before described, a like experiment was performed with the results essentially the same, although differing in the following details: After the great collapse there was a compensatory rise in blood-pressure during the second minute, almost reaching the control height. However, this blood-pressure curve was greatly modified by each respiratory gasp. Respirations were at

first arrested, and later irregular respiratory efforts were noted.

The autopsy immediately after death showed all the venous trunks full, arteries empty, right side of the heart full, left side empty and contracted, the liver much distended.

15. Upon a spaniel, weighing thirty pounds, with control pressure at one hundred and sixty-four millimetres, under ether anæsthesia, the previous technique was repeated. After the first momentary rise in the blood-pressure there was a considerable fall, then great collapse, after which there was a compensatory partial recovery, then a gradual decline until death. The respirations were slow and shallow, afterwards somewhat increased, then they gradually declined until death. At the end of four minutes and ten seconds no further respiratory or cardiac activity could be noticed, although the animal was practically dead within twenty seconds.

Autopsy showed all the veins full, the arteries empty, the blood cyanotic, right side of the heart very full, especially the right auricle. The left side contained a little blood, but was fairly well contracted.

16 and 17. A double experiment, in which the same technique as in the preceding case was carried out. In one animal water at 33° F. and in the other at 180° F. was used. For the cold-water experiment a fourteen-and-one-half-pound spaniel, with control blood-pressure at one hundred and twenty millimetres, under ether, was employed. For the hot-water experiment use was made of a thirteen-and-one-half-pound mongrel, with carotid blood-pressure at one hundred and twenty milli-

metres, also under ether. The animals were both drowned in the same way and at the same time, with the funnels at the same height. In the cold-water experiment there was at first a rapid slight drop in the blood-pressure, followed by "vagal" beats, reaching a considerable height. Then there was a very rapid fall, the heart executing rather faint beats, at fairly regular intervals, during five minutes and twenty seconds. Respiratory efforts were slow and soon failed. In the hot-water experiment the blood-pressure fell immediately to zero, then there were thirteen strong "vagal" beats, a few very weak ones, then death. The last observable action of either the heart or the respiration was at the end of two minutes.

The autopsies in both cases revealed similar conditions, excepting with regard to the contraction of the heart. In the hot-water dog the heart was so firmly contracted as to give the impression that the protoplasm was coagulated.

18 and 19. A double experiment, with the technique as in the preceding case, using in the one dog water at 170° F. and in the other water at 36° F. For the hot-water experiment the animal used was a thirty-seven-pound Scotch collie, with control blood-pressure at one hundred and sixty millimetres. The cold-water dog was a mongrel, weighing twenty-three pounds, with control blood-pressure at one hundred and forty millimetres. Both were under ether anæsthesia. Both were drowned at the same time and by the same method. At the end of two minutes and ten seconds neither cardiac nor respiratory action appeared in the hot-water dog, whilst in the cold-water dog these functions continued for seven

minutes and thirty seconds. In both cases there was the usual great fall in the blood-pressure. In the hot-water dog the fall amounted to a striking collapse, rather more than in the cold-water experiment. However, the fall in each case was very sudden and great.

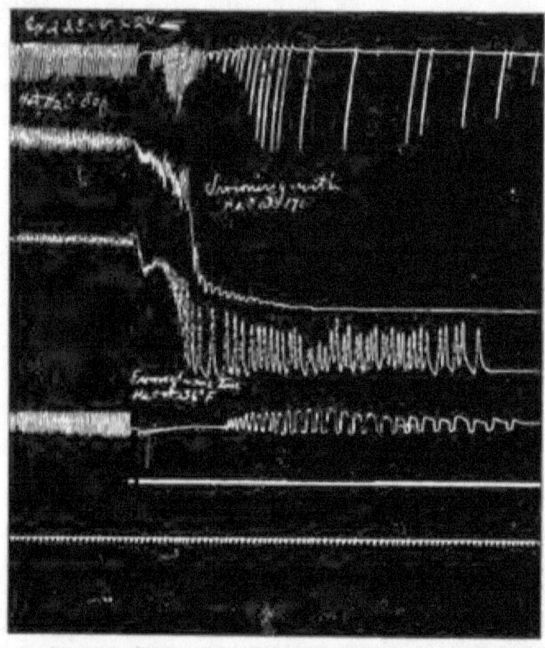

12. DOUBLE DROWNING EXPERIMENT.—The upper tracing represents the respiration in the hot-water (170° F.) experiment, the next tracing the blood-pressure of the same. The third tracing represents the blood-pressure in the cold-water (36° F.) experiment, and the fourth the respiration of the same. The lower two are the signal and the time marks (seconds).

At the autopsies the only difference was that in the hot-water dog the heart was firmly contracted in every chamber; in the cold-water dog the left side was but moderately contracted, while the remainder of the heart was flaccid.

20 and 21. A double experiment, under all the conditions as in the preceding case, the cold-water dog im-

mersed in 33° F. water, the hot-water dog at 170° F. For the cold-water dog, a St. Bernard, weighing sixty-eight pounds, with control blood-pressure at one hundred and forty-four millimetres, was used; for the hot-water dog, a Newfoundland, weighing ninety pounds, with control blood-pressure of one hundred and fifty millimetres. Both animals were drowned simultaneously by like methods. In the cold-water experiment there were more or less respiratory and cardiac efforts for thirteen minutes. In the hot-water experiment the cardiac and the respiratory efforts continued ten minutes. This experiment in almost every respect was similar to the preceding, the greater drop in the blood-pressure appearing in the hot-water dog, with the compensatory rise later. In both cases the circulatory apparatus was practically arrested within the first ten seconds.

At the autopsies the one important difference between the two dogs was the condition of the heart. The right side of the heart in each dog was distended, especially the right auricle. The left side of the heart in the cold-water dog was flaccid, and in the hot-water dog it was firmly contracted. The left auricle also was firmly contracted.

22 and 23. A double experiment with two animals, each weighing sixteen pounds, in good condition, and each having a control blood-pressure at one hundred and thirty millimetres. Both were under ether anæsthesia. Both dogs were drowned with water at the temperature of the body. The second dog, which we shall name the "vagal" dog, was subjected to dissection of both vagi preliminary to the drowning. The usual effect, in-

creased height in blood-pressure, was noted on severing the vagi. At the same time respiratory movements were decreased in frequency and increased in amplitude. Both dogs were drowned at the same time and in the same manner. The animal with intact vagi died displaying almost the precise phenomena described in the preceding cases. The "vagal" dog suffered a fall in blood-pressure quite equal to the control dog and died within two minutes. It was observed, however, that the respirations were not arrested temporarily so markedly as in experiments in which the vagi were intact, but respirations went on for twenty seconds, then gradually faded out.

Autopsy.—The autopsy showed conditions entirely similar to those in the preceding experiments in which water at the temperature of the body had been employed.

24 and 25. Double experiment. The first animal was a control, weighing twenty-six pounds, in good condition, with blood-pressure at one hundred and thirty millimetres. In the second animal a physiological dose of atropine was administered previous to the experiment. The effectiveness of the atropine was tested by stimulation with a Du Bois-Reymond coil. Both animals were then drowned in the same way with the same apparatus. The control dog exhibited nearly all the phenomena that have been heretofore described in similar experiments. In the atropine experiment the collapse of the blood-pressure was not so marked as in the control. The fall of the pressure, however, was very great. The respiratory phenomena in the atropine experiment were similar to those in the control.

The autopsy revealed the auricles of both animals

practically in the same condition as in the preceding cases, excepting in the atropine experiment the blood was more clotted than in the control. The autopsies were made simultaneously.

26 and 27. Double experiment. For a control a mongrel, weighing twenty-six pounds, with a blood-pressure of one hundred and twenty-three millimetres, was used. For the "vagal" a dog, weighing twenty-two pounds, with one hundred and twenty millimetres. Ether was administered in both cases. On severing the vagi the usual phenomena followed. Both dogs were drowned simultaneously. The blood-pressure curve in each case was smaller, although in the "vagal" the heart-beats were more rapid, whilst in the control the heart executed the usual slow, strong beats. The immediate effects upon the respiration differed in these animals. In the control the respirations were immediately arrested, then became slowed, but in the "vagal" dog the respirations were not so markedly interfered with, gradually declined in their amplitude, and later in their frequency, until death.

The autopsy showed the usual conditions.

28 and 29. Double experiment. For a control dog, a Skye terrier, twenty-one pounds in weight, blood-pressure at one hundred and seventy millimetres, was used. For the "vagal" dog, a mongrel, weighing twenty-two pounds, blood-pressure one hundred and forty millimetres, was selected. Both animals were drowned under the same conditions as in the preceding cases, and simultaneously. On severing the vagi the usual phenomena occurred,— namely, a rise in blood-pressure, with increase in the fre-

quency of the heart's action, and slowed, deepened respiratory action. The initial fall in both cases was, in the main, the same; the animals lived about the same length of time, but the respiratory effects differed in this respect, that while in the control dog the respirations were immediately arrested, then later partially resumed, in the "vagal" the respirations were not immediately affected, but gradually declined until death.

Autopsy revealed conditions similar to those observed in the preceding cases, excepting that in the control dog the heart was more distended than in the "vagal."

30 and 31. Double experiment. The first animal was given a physiologic dose of atropine, and in the second both vagi were cut. The animals weighed respectively twenty-one and twenty-two pounds, with blood-pressure at one hundred and ten and one hundred and fifteen millimetres. Both were under ether anæsthesia. Both animals were drowned simultaneously in the same manner. In all the main respects the blood-pressure exhibited like phenomena in both animals; the respirations, however, in the atropine experiment were more directly affected, having been arrested immediately; then very slow, shallow efforts were observed; while in the "vagal" experiment respirations were not immediately affected, but went on for a number of strokes in a rhythmic way, and finally ceased.

The autopsy showed similar conditions in every detail.

In the following experiments the animals were subjected to a stream of water forced into the mouth and through the upper air-passages alone, excluding the trachea and respiratory tract.

SURGERY OF THE RESPIRATORY SYSTEM 57

32. A shepherd-dog, weighing thirty pounds, with blood-pressure at one hundred and forty-two millimetres, was subjected to the following technique: The trachea was dissected out, severed from the larynx, and a breathing apparatus attached. The animal's mouth was then opened and a stream of water was forced through the

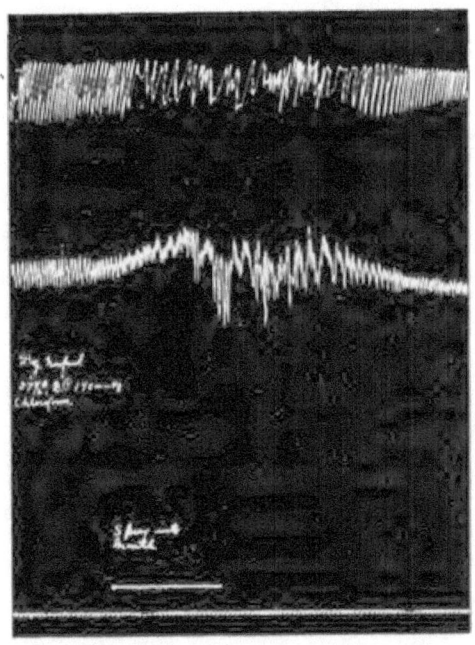

13. EFFECT OF DIRECTING A SPRAY OF WATER FORCIBLY INTO THE PHARYNX WHILE THE TONGUE WAS WELL DEPRESSED.—The upper tracing represents the respiration, the next the blood-pressure, and the lower two the signal and the time-marker respectively.

pharynx into the larynx. The effect upon the blood-pressure was a slight fall, the heart executing extremely long "vagal" strokes; then slow, sweeping beats, characteristic of such action, appeared. The respirations were momentarily arrested, after which they went on undisturbed. This experiment was repeated a number of times, each time producing similar results, though subsequently the results were not so striking. The animal was then killed by pouring water into its trachea through the funnel, as in the preceding cases, exhibiting all the characteristic phenomena.

The observations at the autopsy were the same as in the preceding experiments.

33. The technique of the preceding experiment was carried out in every detail, yielding like results. Water at different temperatures was used. It was found that more striking results were obtained by using hot water than when the water was at the temperature of the body. The greater the force of the stream, the more marked the alteration in the heart-strokes. Both vagi were then severed, and, on repeating the experiment, the heart-beat was not altered.

34. In this experiment the same technique was carried out as in 33, with practically the same results.

Summary.—The experimental evidence obtained would seem to establish an important factor in the effect upon the circulation from mechanic pressure of the column of water in drowning. When the animal was barely submerged in water the effect upon the blood-pressure was not nearly so great, the usual marked collapse did not appear, and the animal lived longer. Not only did the circulatory apparatus maintain its integrity longer, but the respiratory as well. Although not noted in the protocols, in a number of instances the height of the tubing through which the water was poured was varied, and with this variation, as nearly as could be estimated, there was a comparative diminution or increase, as the case might be, in the extent of the collapse in the blood-pressure,—that is to say, in a given case, the higher the column of water, the greater the collapse in the blood-pressure.

Now, throughout the experiments it will be noted that the respiratory action was rather summarily arrested; after a pause, or at least after a period of very slight activity, the respirations became gradually stronger and then faded away; all of which, as a rule, occurred during the period of from thirty seconds to several minutes. The circulation was in nearly every experiment wholly overcome within fifteen seconds, and in a goodly number of cases within from five to ten seconds. The collapse was most marked during the inspiratory phase of respiration. The effect of respiratory action upon the blood-pressure was most marked, producing, as it did, extreme variations in the blood-pressure curve. These variations seemed to be rather out of proportion to the force of the expiratory action.

Now, in what respect do these phenomena correspond with the phenomena of asphyxia? In asphyxia there is not a sudden cessation of either the respiratory or the cardiac activity. When the exchange of air is suddenly arrested, as by clamping the trachea, respirations may momentarily cease, but this cessation is only momentary, and respiratory efforts are carried on with increasing vigor for a considerable length of time. The respiratory curve is far greater and more vigorous in an asphyxia experiment than in the drowning experiments.

As to the circulatory phenomena: In asphyxia there may be a temporary decline—usually a slight decline—in the blood-pressure. This decline is very quickly recovered from, and there is an actual rise in the blood-pressure above the level at which it was before the asphyxia was induced. The heart continues to beat in asphyxia for as long as from three to twelve minutes. In fresh animals, as were those in which drowning experiments were performed, the blood-pressure in asphyxia never underwent this fall; on the contrary, as it was said before, in asphyxia it rose. The characteristics of the blood-pressure curve in asphyxia are an immediate gradual temporary decline, then a gradual rise above the height it was before the experiment; late in the experiment, there will appear Traube-Hering curves, and finally the heart ceases beating in diastole. It is true that in producing asphyxia, either by drowning or by closing the trachea, there is no exchange of gases in the lungs. There is an immediate cutting off of the supply of oxygen. On what grounds, then, may the very great difference between the phenomena of these two conditions be explained? It cannot be a reflex action through the vagi, producing a powerful inhibition upon the heart and upon the respirations; for in the experiments in which the vagi were cut, preliminary to the drowning, as striking a collapse in the blood-pressure occurred as before. This, in itself, is conclusive so far as the trunks of the vagi are concerned; but it may be argued that some influence upon the terminals of the vagi may cause inhibition of the heart. In a series of experiments in which atropine was given previous to the experiment, in dosage sufficient to paralyze the nerve-endings in the heart, as proved by laying the electrode of a Du Bois-Reymond apparatus upon the vagus, producing no inhibition while the secondary coil was lapping the primary by one-half, when the usual drowning experiment was then performed, the collapse in blood-pressure occurred the same as in the cases in which the animal had not received this drug. There is no other mechanism, then, through which afferent impulses may produce an inhibitory action upon the heart, producing thereby such staggering fall in blood-pressure; and, besides, the character of the heart-beats at the moment of the great staggering fall in blood-pressure is not that of a "vagal" beat. Then is it possible that the presence of water in the lungs may cause a stimulation of vasomotor nerves producing these effects? In the first place, it is impossible to produce so abrupt and staggering a fall by acting upon all the vaso-motors of the body at once. Much less would it be possible to account for such fall in the blood-pressure by supposing the vaso-motors supplying the blood-vessels in the lungs to be stimulated. At best the vasomotors supplying the lungs are not capable of producing great alterations in the blood-pressure.

Then, may it be due to the temperature of the water? A series of double experiments was made in which the water introduced into one animal was hot, while another animal was drowned simultaneously with cold water; the fall in both cases was practically the same. The animal taking the hot water showed total loss of cardiac and respiratory activity earlier than the animal taking the cold water. However, the feeble efforts after the first fifteen seconds in either case were not sufficient to re-establish either the circulation or the respiration. In a further series of cases, in which the water was introduced at the temperature of the body, the phenomena did not essentially differ from the phenomena in the experiments in which the water was used either hot or cold. While in some respects the temperature of the water had characteristic effects, these characteristic effects were more marked in the case of hot water, and were shown on necropsy in the firm contraction of the heart. The left auricle was most firmly contracted, and the left ventricle next.

Then, if this striking collapse in the blood-pressure is not due to the temperature of the water, to a vasomotor disturbance, nor to an excitation of the vagi or any of their terminals, nor to the asphyxia effects, to what may these phenomena be attributed? Before proposing an explanation it would be well to call attention to some estimation of the comparative value of this staggering fall in the blood-pressure. In order to make a comparative estimate the superior *vena cava* and inferior *vena cava* were severed, letting free torrents of blood, producing thereby as staggering a fall in blood-pressure as is possible to do by hemorrhage. Even when severing the one and leaving the other intact, the collapse in blood-pressure was not so great as that in the drowning experiments.

Whatever may be the mechanism of this sudden collapse in drowning, it must, to a very extraordinary degree, block the entire circulation. Now, the circulation in the lungs is not under a high pressure; especially in the capillaries, there is not much resistance to overcome in this circulation. Without undertaking to state the exact pressure, it was calculated that in all the experiments in which there was a collapse in the blood-pressure, the column of water in the tube in the apparatus exerted a pressure greater than the capillary pressure in the lungs. So that as long as the apparatus was filled with water a pressure was exerted indirectly, through the medium of the air filling the alveoli of the lungs, upon all the capillaries, and, this pressure being greater than the capillary pressure, the entire pulmonary circulation was blocked. The momentary rise in the blood-pressure at the beginning of the drowning was due to the fact that the pressure produced by the water was greater than that in the capillaries; the blood in the pulmonary capillaries was forced out, and thereby increasing the quantity of blood to the heart and momentarily raising the pressure. The force sufficient to drive out this blood was sufficient to block the further flow of blood in the capillaries, and therefore the column of water in the apparatus at such a height above the lungs was as effective a block to the circulation as a ligature.

This view of the cause of the collapse of the circulation was supported by every necropsy made. In every case the right auricle was enormously distended, the right ventricle less so, while the left side of the heart was empty. It is apparent that the right side of the heart was unable to force the blood in its chamber through the lungs against this superior resistance. The necropsies also revealed enormous distention of all the venous trunks. The right ventricle, being unable to empty itself, could not receive the blood from the right auricle, and, it being unable to empty itself, could not receive the blood from the large venous trunks, and they being unable to empty themselves could not take the blood from the smaller venous trunks, thus leading to their distention. The arteries were empty because no blood had reached the left side of the heart from the lungs.

As to the respirations: It was noted in the protocols that in the experiments in which a preliminary dose of atropine had been given the respirations were interfered with more markedly than in the case in which the vagi had been severed. The same comparison was made between the control experiments and the "vagal" experiments,—namely, that in the "vagal" experiments the respirations were not so much interfered with as in the control experiments. This is interpreted in the following manner: The entrance of water into the respiratory tract set up afferent impulses which were carried up to the respiratory centre through the vagi; having reached the respiratory centre, afferent impulses were sent out, causing interference with the respirations. Now, in the cases in which the vagi had been severed, the afferent impulses towards the respiratory centre had been interrupted, the respiratory centre could not receive these afferent impulses, and for this reason the respirations were not affected.

It was stated that the respiratory action in these experiments was much less active than in cases of death by asphyxia. This difference between the respective respiratory actions is accounted for as follows: In the first place, what has just been stated as to the difference between the respiratory action when the vagi are cut and when they are intact tends to show that there is a reflex inhibition of the respiratory action in drowning. This, however, does not sufficiently account for the difference between the respiratory phenomena in drowning and in asphyxia, for in the experiments in which the vagi had been severed before the drowning, the respiratory action was less than in typical cases of asphyxia. The reason why this respiratory effort is less in this class of cases of drowning than in asphyxia is, probably, on account of the very great fall in the blood-pressure, which occurs at the onset of the drowning. That the respiratory action is markedly diminished by a lowering of the blood-pressure in the brain has been well established, so that the great collapse in the blood-pressure itself in these experiments would be sufficient markedly to diminish the respiratory action. No such collapse occurred in asphyxia experiments. That the collapse in the blood-pressure has an effect upon the respiratory action is directly shown in the experiments in which the drowning pressure was but slight, producing thereby but little fall in the blood-pressure; in these cases the respiratory action was decidedly greater than in the cases in which there was a marked collapse in the blood-pressure.

The series of experiments in which water was thrown into the pharynx and against the opening of the larynx, while the remainder of the pulmonary tract was excluded, producing thereby a marked inhibitory effect upon the heart and the respiration, tends to introduce another factor in the sudden collapse in drowning. If the foregoing deductions are correct, the circulatory factor in drowning in cases in which the column of water exerts its pressure upon the pulmonary tract is fully as potent as the respiratory.

A RESEARCH INTO THE CAUSE OF CERTAIN SYMPTOMS OBSERVED ON ENTERING AN ATMOSPHERE OF INCREASED BAROMETRIC PRESSURE

Preliminary Remarks.—On entering an atmosphere of increased barometric pressure most persons experience some peculiar symptoms, such as oppression, dizziness, occasional fainting, difficult breathing, very rapid pulse, etc. The phenomena observed in the research on drowning suggested that some light might be thrown upon this subject by experiments made along similar lines.

Protocols.—The following technique was employed in the experiments: The animals were placed under full ether anæsthesia, the trachea dissected out, and a canula tied firmly into it; a heavy tubing was then tied upon the canula, and this was connected with a strong, leather bellows, from which a large quantity of air might be suddenly or slowly driven into the lungs,—that is to say, the intra-broncho-pulmonary pressure might be increased, as would be the case in increased barometric pressure. The blood-pressure was taken in the carotid artery. The respirations were neglected, as they could express nothing in such experiments.

1. A fox-terrier, weighing twenty pounds, in good condition, with an initial blood-pressure of one hundred and forty-six millimetres, was subjected to the experiment outlined. The bellows was suddenly emptied into the pulmonary tract of the dog, producing a very great immediate fall in the blood-pressure. The rubber tubing at the close of the injection of the air was clamped, the bellows refilled; then, as the air was again passing into

the tube, it was unclamped, and more air forced into the lungs. In this manner the blood-pressure fell abruptly

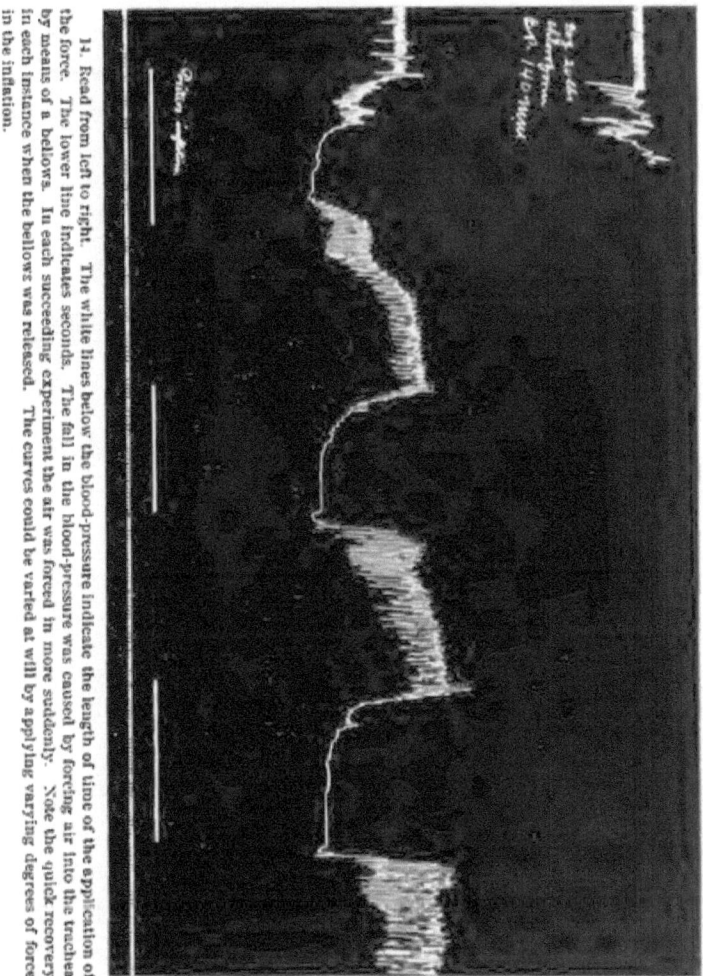

14. Read from left to right. The white lines below the blood-pressure indicate the length of time of the application of the force. The lower line indicates seconds. The fall in the blood-pressure was caused by forcing air into the trachea by means of a bellows. In each succeeding experiment the air was forced in more suddenly. Note the quick recovery in each instance when the bellows was released. The curves could be varied at will by applying varying degrees of force in the inflation.

to the abscissa line. The animal was killed almost instantly thereby.

An immediate autopsy showed the right heart to be engorged, the left heart empty.

68   EXPERIMENTAL RESEARCH INTO THE

2. A spaniel, weighing twenty-four pounds, was subjected to an experiment similar to the preceding; the blood-pressure suffered a staggering fall, as before. In this instance the pressure fell at the first blast almost to the abscissa line. On removing the clip and allowing the air to escape from the lungs, the blood-pressure leaped up very rapidly, and soon regained its normal height. This

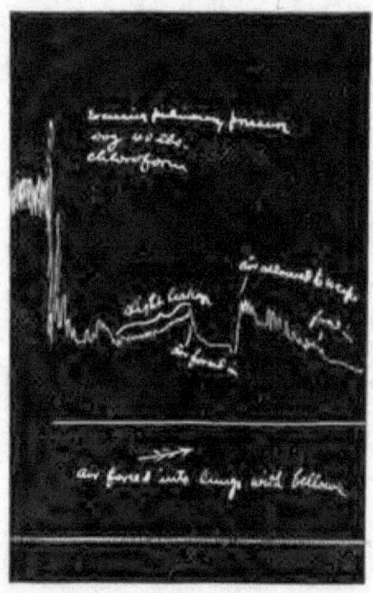

15. POSITIVE INFLATION OF THE LUNGS INCREASING THE TRACHEA-PULMONARY PRESSURE BY MEANS OF A BELLOWS.—Note the momentary rise in the pressure, followed by an immediate staggering fall. The lower line indicates seconds.

was several times repeated, so that there could be no doubt as to the striking effect of an increased intrathoracic pressure upon the circulation. The animal was then killed by again increasing the intrathoracic pressure, so as to block the circulation through the lungs.

The autopsy revealed conditions similar to those observed in the preceding case.

3. A bull-dog, weighing forty pounds, was subjected to an experiment similar to the preceding, with practically the same results. The animal was killed almost instantly. Observations at the autopsy were similar to the preceding.

4. In a strong shepherd-dog a control was made by several vigorous blasts from the bellows, producing collapse in the blood-pressure. The air was immediately allowed to escape, and the blood-pressure mounted up again to the point previously occupied. Then, passing a scalpel between the ribs and severing the descending vena cava, the blood gushed in torrents out of this vessel. The blood-pressure sank, but it sank more gradually than in the case of increased intrapulmonary pressure produced immediately before by the injection of air.

5. A large St. Bernard was subjected to the same treatment as in the foregoing case, producing first the collapse in the blood-pressure by forcing air into the lungs; then, allowing the blood-pressure to recover itself, the inferior vena cava was severed and the blood permitted to gush forth freely. A great fall in blood-pressure was produced, but this fall was not so sudden, nor was the collapse so marked as in the experiment of forcing in air.

Summary of Experimental Evidence.—The evidence of these experiments and of the experiments on drowning tends to show that a sufficient increase in the intrapulmonary pressure may produce a collapse of the circulation. It robs the left heart directly of blood, and therefore causes a greater collapse than can be produced by severing either the superior or the inferior vena cava alone.

Now, as a workman enters a tunnel under high barometric pressure the respiratory tract is subjected to an increase in pressure, and the immediate symptoms may be interpreted from this main fact. The dizziness, the difficult breathing, the soft, rapid pulse, all would be produced by a great fall in the blood-pressure. The respiratory symptoms may be produced by a sudden fall in the blood-pressure as well as diminished circulation of the blood, thereby diminishing the exchange of gas in the lungs, which, in itself, is sufficient to cause an increased respiratory action and produce the respiratory distress.

FOREIGN BODIES IN THE PHARYNX AND ŒSOPHAGUS

Preliminary Remarks.—The symptoms produced by foreign bodies of considerable size lodged in the pharynx and certain portions of the œsophagus bear a close resemblance to those occasioned by the presence of foreign bodies in certain parts of the respiratory tract.

Among the symptoms of choking may be mentioned cyanosis, asphyxia, collapse, and slow pulse; these symptoms so closely resemble those produced by foreign bodies in the larynx that the one condition has been frequently mistaken for the other, and in many cases the differential diagnosis from the subjective symptoms alone has been quite impossible.

The following research was undertaken, therefore, to attempt an explanation of these phenomena:

Protocols.—1. On a mongrel dog weighing twenty-eight pounds, whose blood-pressure registered one hundred and thirty-four millimetres, under ether anæsthesia, the following experiments were performed:

(a) The œsophagus was exposed by careful dissection, so as to avoid severing any of the nerves. This tube was then opened by a longitudinal slit in its middle, and the portions immediately above and below this slit were subjected to dilatation by means of a large bulb and also with a uterine dilator. Although the dilatation was forcible, no effect upon blood-pressure was produced. It is doubtful whether the slight slowing of the respiration was a coincidence or an effect.

(b) Then, passing the instruments down to a point opposite the bifurcation of the trachea and producing there a similar dilatation, like observations were made.

(c) The instruments were passed upward to a point opposite the larynx and dilatation was made; there was a slight rise in blood-pressure and a temporary arrest in respiration occurred, this temporary arrest being followed by slow respiratory efforts during the remainder of the dilatation.

16. EXPERIMENT ON CHOKING. BY PRESSING A GLOBULAR FOREIGN BODY INTO THE ŒSOPHAGUS OPPOSITE THE LARYNX.— Note the temporary inhibition of the respiration (upper tracing) and the staggering fall in the blood-pressure (second tracing).

(d) The same experiment was repeated with similar results. Then, passing the dilator down to the cardiac orifice of the œsophagus and dilating this tube, there was produced a slight fall of blood-pressure, and the respirations were diminished in frequency.

(e) Introducing from below upward the rounded handle of an instrument larger than the opening into the pharynx, producing thereby as great a dilatation as possible of the upper portion of the œsophagus, there followed a very marked fall in blood-pressure, and the character of the heart-beats displayed the distinct effect of "vagal" action. The respirations at the same time were very much slowed. The middle portion of the œsophagus was then again dilated with as much force as possible, resulting in a slight irregularity of respiration with an unchanged blood-pressure curve.

(f) Forcible dilatation of the œsophagus again upon a level with the upper portion of the larynx and at the end of the pharynx produced a very great immediate fall of the blood-pressure and complete arrest of respiration. The narrowest portions of the œsophageal tube were at the pharyngeal and the cardiac extremities.

2. On a mongrel dog, in poor condition, weighing twenty-three pounds, under ether anæsthesia, the following experiments were performed:

(a) Introduction of the rounded, wooden handle of an instrument into the pharynx produced a marked "vagal" cardiac action and irregular, slowed respirations. This was twice repeated, with the same result each time. The middle portion of the œsophagus was again dilated, and it was found that but minor effects were produced from as little or as much dilatation as was possible of this tube from the lower level of the larynx down to near the cardiac orifice.

(b) On forcing an obturator into the œsophagus very slowly upward until it entered the cavity of the mouth, no appreciable effects were produced until it reached a point opposite the upper half of the larynx, when a marked fall in the blood-pressure occurred and temporary respiratory arrest.

(c) An injection of atropine, in sufficient dose (as proven by the DuBois-Reymond stimulation) to paralyze the nerve-endings of the vagi, was given. Then, on repeating the dilatation or choking in the area opposite the upper portion of the larynx and in the pharynx, no effect upon the blood-pressure was observed; the respirations were arrested as before.

3. On a Newfoundland dog weighing forty pounds, in good condition, with a blood-pressure of one hundred and twenty-eight millimetres, under ether anæsthesia, the following experiments were performed:

(a) The rounded wooden end of an instrument, larger than the normal opening of the pharynx, was rather forcibly pressed through the mouth into the pharynx. This produced an immediate and very considerable fall in blood-pressure with temporary arrest of respiration, followed by irregular, slow, respiratory efforts. Introducing the same into the cardiac end of the œsophagus through an incision in the stomach, there was a slight fall in blood-pressure and slightly irregular respirations. Then, by exposing the œsophagus in the middle of the neck and making a longitudinal opening, the instrument was passed through this incision upward, forcibly dilating at a point opposite the larynx, whereupon there was an immediate and very considerable fall in blood-pressure, with temporary arrest of the heart and of respiration.

(b) A physiologic dose of atropine was given, and its

efficiency proved in the usual way. Then a repetition of the foregoing experiment produced but little effect upon the blood-pressure, the little being sometimes a rise, sometimes a fall ; but it did not prevent the arrest of respiration.

4. On a bull-dog in good condition, weighing twenty-eight pounds, having a blood-pressure of one hundred and forty-five millimetres, under ether anæsthesia, the following experiments were performed :

(*a*) Forcible dilatation of the œsophagus in its middle portion produced a rise in blood-pressure and a slight slowing of respiration. This was repeated twice, in different portions of the œsophagus, with like results.

(*b*) The same experiment opposite the upper portion of the larynx produced a marked fall in blood-pressure, with " vagal " beats and temporary arrest of respiration ; later on the respirations recovered their normal rhythm.

(*c*) A physiologic dose of atropine, proved in the usual way, prevented the effect upon the heart, as was shown in the preceding case, but the respirations were arrested again, as before. Now, making sections of both superior laryngeal nerves and repeating the experiment of choking opposite the larynx, there was no effect produced upon either the cardiac or the respiratory action.

Traction downward upon the œsophagus produced a fall in the blood-pressure, but the characteristic " vagal " beats were not observed and the respirations were unaltered. Traction upward produced a slight rise in blood-pressure, with no alteration in respiration.

5. On a mongrel dog weighing twenty pounds, blood-pressure one hundred and forty-eight millimetres, in good

health, under chloroform anæsthesia, the following experiments were performed:

(a) The handle of a chisel passed into the pharynx and forced down in imitation of choking produced a marked fall in the blood-pressure and arrest of respiration. The pressure of the chisel did not come in contact with the larynx, but with the pharynx and the base of the tongue.

(b) A physiologic dose of atropine prevented the appearance of the blood-pressure phenomena on repetition of the experiments. The respirations were arrested as before. Various portions of the œsophagus were then tested, and with the exception of the cardiac end, there was but little effect upon either the respiration or the blood-pressure. The cardiac end was finally dilated with a Goodell dilator with very great force, the operation resulting in a slight slowing of respiration. Finally, on severing both superior laryngeal nerves, then repeating the pharyngeal choking experiment, there was a rise produced in blood-pressure instead of a fall, and the respirations remained undisturbed. Downward traction with considerable force upon the œsophagus produced a fall in blood-pressure with slight respiratory slowing; upward traction produced a much less fall in blood-pressure and no respiratory alteration.

6. On a bird-dog weighing thirty-two and one-half pounds, with blood-pressure of ninety millimetres, in good condition, under chloroform anæsthesia, the following experiments were performed:

(a) The wooden handle of an instrument was placed deep in the pharynx, and the free end carried outward into the angle of the mouth, so as to produce a severe prying pressure against the walls of the pharynx. This produced a staggering fall in the blood-pressure, with complete temporary arrest of the heart and of the respiration. On severing both superior laryngeal nerves and repeating the experiment there was a very marked fall in the blood-pressure, with temporary arrest of the heart as before, but the *respirations remained unaltered.*

(*b*) Division of the hypoglossal nerves on both sides and repeating the experiment produced a temporary arrest of the heart, with a staggering fall of the blood-pressure, but no effect upon the respiration.

(*c*) As a control, very severe intralaryngeal manipulation was made, which produced no effect upon either the respiration or the heart's action. Then choking was again done in the same violent way, through the mouth with the wooden handle of an instrument, producing thereby the same inhibition of the heart as before.

7. On a thirty-nine-pound dog, with blood-pressure at one hundred and ten millimetres, in poor condition, under chloroform anæsthesia, the following experiments were performed: The glossopharyngeal, the hypoglossal, and the superior laryngeal nerves on each side were exposed, looped with thread, and the animal prepared for the following tests:

(*a*) As a control, the animal was choked from below upward, opposite the upper portion of the larynx, with the production of the usual inhibitory phenomena; then, through the mouth with the usual force, pharyngeal choking was practised, and resulted in the usual inhibitory phenomena.

(b) Now both glossopharyngeal nerves were severed. Then, on repeating the above control manipulations, like inhibitory phenomena were produced.

(c) Division of both hypoglossal nerves and then repeating the choking experiments produced practically the same reflex phenomena upon both respiration and the heart.

(d) An injection of one-one-hundredth of a grain of atropine into the jugular vein. Repetition of the same experiments produced no effect upon either the respiration or the heart's action. In the experiment of choking, especially when the foreign body was large, there was evidence that the vagus itself was mechanically stimulated, either directly by a direct continuity of pressure, or indirectly by being dragged on by structures that had a connecting anatomical relation with it, and that thereby a direct inhibition was produced.

Summary of Experimental Evidence.—It is noted in all the cases that the narrowest portions of the pharyngo-œsophageal tract were at the pharyngeal and the gastric ends of the tube respectively. Artificial choking produced at the gastric end of the œsophagus had some effect upon the respiration, usually slowing it, and caused, also, some fall in the blood-pressure, but in no case were the effects of a striking nature. They were properly classed as minor phenomena. In no experiment on any part of the œsophageal tract up to a point opposite the larynx were more than minor effects produced. Severe dilatation of this portion of the tube sometimes produced a rise, sometimes a fall, in the blood-pressure; the respiratory effects were never of much importance. However, at the points opposite the larynx, especially the upper portion of the larynx, very marked respiratory and circulatory phenomena were caused by artificial choking. These effects were doubtless inhibitory and through the vagus.

That the cardiac action was through the vagus is proved by the fact that physiologic dosage of atropine in every case prevented the fall in blood-pressure and the "vagal" strokes. That the superior laryngeal nerve was not the only path over which the inhibiting impulses passed is proved by the fact that the pharyngeal choking, especially when accompanied by considerable violence, produced a marked fall—in fact, a collapse—in blood-pressure after the superior laryngeal had been severed. That it was not through the hypoglossal or the glossopharyngeal that this collapse was produced was proved by the observation that the same phenomena occurred after their severance.

Direct observation of the mechanic dragging upon the upper portion of the vagus through the medium of the numerous anatomic structures lying in such close apposition to it and so closely interwoven, together with the negative proof adduced by the experiments, was sufficient to show that a part at least of the inhibition phenomena were due to a mechanic stimulation of the vagus itself. It is highly probable that the marked respiratory arrest was due to a mechanic irritation of the superior laryngeal; but that it was not the only tract over which impulses passed was proved by a slight alteration in the respiratory action after the superior laryngeal had been severed, although no such marked effects as before could be produced. It is true, however, that all the evidence goes to show that the superior laryngeal is the only source through which choking may cause very striking reflex inhibition of the respiration. Tolerance, so far as the respirations are concerned, is very soon acquired; it is later acquired on the part of the circulation.

Some Observations.—Choking, then, produced symptoms of reflex inhibition, partly through the superior laryngeal, and partly through the trunk of the vagus itself, symptoms almost identical with those of severe irritation of the larynx itself. In the case of the larynx all the symptoms are due to mechanic stimulation of the superior laryngeal nerves. In a given case, then, of threatened asphyxia, if operative procedures are to be undertaken for dislodgement of a foreign body, it would be well to be guided by the same principles that guided in the treatment of foreign bodies in the larynx, at least so far as to give the preliminary dosage of atropine to prevent a great collapse or possible death during the operative procedures. It is perfectly apparent that in case there is a history of choking without inhibition phenomena, the foreign body, if large enough to cause much pressure, must be at a point below the level of the larynx.

FOREIGN BODIES IN THE TRACHEA AND THE LARYNX

Preliminary Remarks.—In cases of foreign bodies in the respiratory tract there is often from the symptoms alone great difficulty in determining their location. It is difficult also to decide whether or not an operation is indicated, and, if indicated, where this operation is to be performed. The number of futile attempts at finding foreign bodies, and the number of catastrophes that have followed such accidents, suggested the following research.

Protocols.—The following experiments were made with a view to ascertaining the comparative values of *chloroform* and *ether* in point of safety in operations involving the upper air-passages :

1. On a sixteen-pound mongrel, in fair condition, under *ether* anæsthesia, the following dissections were made: Both vagi were dissected free from the carotids throughout the length of the neck; then the superior laryngeals were dissected out from their origin in the vagi to their entrance into the box of the larynx; next the glossopharyngeal was dissected out from its exit from the cranial cavity to its terminals in the muscles; finally, the hypoglossal was similarly dissected, thus exposing these structures on both sides. During the dissections the animal breathed fairly well without requiring much attention to the ether anæsthesia, which was administered by means of a tracheal tube, to which a rubber tube and funnel were attached, and into the latter cotton-wool was thrust. Manipulations of these nerves, especially of the vagus and inferior laryngeal, interfered with the respiratory action as well as the circulatory. However, in no instance was it necessary to supply artificial respiration.

2. On a seventeen-pound bird-dog, under *chloroform* anæsthesia, dissections were carried out as nearly like the preceding as possible, occupying as nearly as possible the same length of time in their performance. During the anæsthesia it was noted that the blood-pressure would sink whenever the respirations became slowed, and that

the pressure would recover itself upon removal of the anæsthetic. Early in the experiment it was found necessary to maintain artificial respiration for a short period of time, and during the latter part of the dissection respirations became extremely shallow and weak. Manipulation of the superior laryngeal or the vagus interfered with the respirations more markedly than in the preceding experiment.

3. On a water-spaniel weighing twenty-four pounds and in good condition, under *ether* anæsthesia, the following experiment was performed: The trachea and the larynx were both dissected free, the trachea severed at the cricoid junction, then the superior laryngeal was dissected out from its origin to its termination on both sides of the neck. A finger was passed up within the larynx, producing thereby an irritation in the area for inhibition. The superior laryngeals were stimulated by the faradic current, producing temporary arrest of respiration and complete inhibition of the heart. On cessation of the stimulation both the cardiac and the respiratory action were resumed. After this the animal was allowed to rest undisturbed with these structures exposed to the air. The respirations were decreased in amplitude and in frequency,

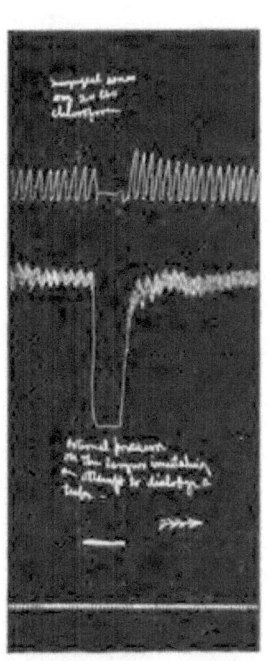

17. EXPERIMENT ON THE REMOVAL OF A LARYNGEAL TUBE BY MEANS OF EXTERNAL MANIPCLATION.—The upper tracing represents the respiration, the next the blood-pressure, below which are the signal and the time.

but at no time ceased, excepting when some special irritation produced reflex inhibition.

4. A bull-terrier, weighing twenty-eight pounds, in fair condition, under *chloroform* anæsthesia, was subjected to as nearly as possible the same procedures as in the foregoing case, occupying about the same length of time, with the following result: The reflex inhibitions on both the respiration and the heart were under chloroform anæsthesia comparatively more profound. During the dissection respirations were so shallow as to cause considerable cyanosis, and for a time artificial respirations were supplied. At the close of the experiment, when the animal was allowed to rest, the respiratory action was extremely feeble and the blood-pressure had fallen more than one-half from its original height.

5. On a mongrel cur, weighing eighteen pounds, under *ether* anæsthesia, the following experiments were performed: The superior laryngeal nerves on both sides were exposed and subjected to stimulation by mechanical manipulation, producing thereby a reflex arrest of respiration. The faradic current was then applied, producing reflex inhibition of both the respiratory and the cardiac action. This was repeated twice on each side. The normal action was regained with a fair degree of readiness after cessation of the stimulation in each instance. Artificial respirations were not required.

6. On a water-spaniel weighing twenty-two pounds, in fair condition, under *chloroform* anæsthesia, during the technique of dissecting out the trachea and applying the tracheal apparatus for maintaining anæsthesia, respirations ceased and artificial respiration was supplied during

a period of five minutes, after which very slow, voluntary efforts were noted, gradually increasing in amplitude and rapidity until an over-respiratory action and recovery occurred. On reducing the animal to surgical anæsthesia again, a like result was observed. During the experiment, carried out as nearly as possible as in the preceding case, the respirations were very much slowed and quite shallow, so that it was difficult to obtain a good tracing. In these experiments the larynx was exposed to all kinds of rude manipulation, for the purpose of facilitating a comparison between the results under chloroform and under ether anæsthesia.

7. This experiment and the following were performed for the purpose of determining whether or not traction on the trachea from the lungs upward, or from the larynx downward, or lifting the entire trachea forcibly out of its bed, caused an effect upon the blood-pressure and the respiration.

On a mongrel dog weighing twenty pounds, under ether anæsthesia, the following experiment was performed: The trachea was freely exposed by bloodless dissection, and after securing a control tracing of the respiration and the blood-pressure, a hook was placed between the rings of the upper portion of the trachea and the latter organ was pulled strongly upward from the lungs in the line of its own axis. This produced no effect upon the respiratory rhythm, but it caused a fall in blood-pressure, this fall taking the form of a rather sharply rounded curve, from which it quickly recovered. Then, repeating the same manœuvre, like effects were produced. The line of traction was then reversed and the trachea dragged downward on the line of its own axis, producing a dragging upon the entire larynx and the pharynx. This was followed by a decided fall in the blood-pressure and a temporary slowing of respiration. This experiment was twice repeated, with like results, though in each subsequent repetition the phenomena were not so marked as in the one immediately preceding. In dragging upward on the trachea, when the dragging was quickly applied, the fall in blood-pressure was correspondingly more abrupt, and suggested the probable action of mechanical causes interfering with the return circulation in the venous trunks which pass in such close anatomical relation to the trachea.

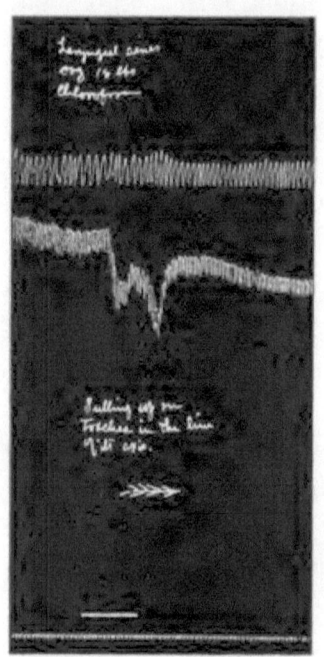

18. EXPERIMENT ON THE TRACHEA.—The upper tracing represents respiration, the next the blood-pressure, and the lower two the signal and the time (in seconds).

8. The animal, a fox-terrier, weighing twenty pounds, was subjected to experiments similar to those in the preceding case. The phenomena attending upward traction on the trachea were practically the same as in the preceding case, except that there seemed to be some increase in the depth of respirations. On dragging down on the trachea the respirations were more decidedly slowed, and for a short period at the first dragging, respirations were

arrested. On raising the trachea up out of its bed with some force, results similar to dragging down upon the trachea were produced.

9. On a bull-dog weighing forty pounds, under ether anæsthesia, experiments similar to the preceding were performed, with the same results, except that in dragging forcibly down upon the trachea there was a distinct inhibitory action upon the heart, whereby sweeping slow beats were executed during the period of most forcible traction. Raising the trachea up from its bed with considerable force produced similar symptoms, though less marked.

While the first six experiments involved the mere dissection and thorough stimulation of certain nerves, they confirm what has been very frequently observed in other experiments upon the respiratory tract,—viz., that animals under chloroform anæsthesia are decidedly more liable to respiratory failure than animals under ether anæsthesia. While it is recognized that it is very difficult to make satisfactory comparative observations, it has seemed that reflex inhibitory phenomena are more profound under chloroform anæsthesia than under ether anæsthesia, inasmuch as in other experimental work, not relating to this present object and purpose, similar observations have been made. These additional experiments have been made principally to test the effect of dissections upon, and about the larynx and trachea, and involving the nerve-trunks supplying this area.

The Physiologic Principles involved.—The physiologic principles involved in this question are practically the same as those indicated under several other headings in this paper,—namely, that no amount of irritation of the mucosa or structure of the trachea, even the cricoid portion of the larynx, is capable of producing any marked sudden effects upon either the circulation or the respiration.

Such an irritation, however, if applied to the middle or upper portion of the larynx, would cause most pronounced reflex inhibitory phenomena. The only reflex phenomenon a foreign body below the larynx could produce is that of coughing. While in the larynx not only coughing may be produced on slight irritation, but, on a greater irritation, reflex arrest of respiration will occur, and on a still greater irritation, in addition to the reflex arrest of respiration, there may be reflex arrest of the heart's action, causing thereby a collapse more or less profound. The symptoms of asphyxia from obstruction need never be mistaken for those of reflex inhibition from the laryngeal area.

In the case of asphyxia from obstruction, the respiratory, as well as the circulatory, activity will continue. The pulse will be strong and full, and slow, until finally, after five to eight minutes, there will be intermissions, then irregular actions, then intermissions, until finally the heart ceases in diastole. In reflex inhibition there is sudden collapse, with sudden cessation of respiratory action, and, if the circulatory action is also affected, there will be no arrest of the heart as sudden as the arrest of the respiration.

Differential Diagnosis between Lodgement in the Trachea and in the Larynx.—A foreign body lodging in the trachea and not dislodged from the trachea during the attacks of coughing, is incapable of producing any effect other than that indirectly produced by the powerful respiratory alterations in the act of coughing. If the foreign body is so large and the swelling so great as to completely obstruct the tube, the symptoms of asphyxia, as above described, will appear. If, however, the foreign body is lodged in the larynx, there are likely to appear attacks of cyanosis, sudden in onset, during which the subject becomes cyanotic, greatly depressed,—in short, passes into collapse,—and, after a variable period, respirations gradually are resumed, the cyanosis disappears, and the heart again beats normally for that particular case.

A slight movement, or a cough, or an attempt at examination, may precipitate such a collapse. Furthermore, supposing a foreign body to have been inspired into the trachea and during a violent fit of coughing to have been expelled from the trachea into the larynx, such reflex inhibition would most likely suddenly appear. If, then, a sudden collapse takes its onset in the middle or the latter part of a severe coughing spell, it may be assumed that that foreign body had not been in the larynx at the beginning of the coughing, but it was forced into the larynx by the powerful expiratory efforts.

Preliminary Preparation for Extraction of the Foreign Body.—In all cases of operative procedure to extract foreign bodies from the respiratory tract it would seem a judicious precaution to inject a sufficient dosage of atropine to protect the heart against the reflex inhibitory impulses through the superior laryngeal nerve, so that, if respirations should fail, the circulation will be sufficiently guarded, so that, while artificial respirations are maintained, the operation may be completed. This is especially necessary in cases in which the diagnosis has been made of lodgement in the larynx.

However, in cases not so diagnosticated, there is by no means a certainty that in the fits of coughing the foreign body will not be forced into that organ before the technique can be completed. In cases in which it is practicable, a spray of cocaine or an intralaryngeal application of cocaine are of the greatest service in protecting the patient against sudden collapse. As has been amply proved, experimentally as well as clinically, the local application of cocaine upon the laryngeal mucosa, even with so weak a solution as one-half per cent., completely prevents the reflex inhibitory effect upon both the respiration and the heart, no matter to what extent the irritation of the larynx may be carried. In cases in which this preliminary preparation can be satisfactorily carried out, it would give an absolutely certain precaution against sudden collapse or death.

On the Technique of the Operative Procedure.—In every case in which it is possible, local rather than general anæsthesia should be employed. The principal reasons for this preference are the following: If the subject can be controlled so that the operation may be performed under local anæsthesia, there is less disturbance, less struggle, and therefore less opportunity for dislodging the foreign body; the patient may aid the surgeon in various ways in his attempt to recover the foreign body; and finally, when the respiratory passages are so occluded as to require the action of extraordinary muscles of respiration it would be almost fatal to attempt a general anæsthetic, because in this case the exchange of air can only be accomplished by the aid of the extraordinary muscles. Now, a general anæsthetic paralyzes these extraordinary muscles of respiration, and therefore the burden of supplying air is thrown upon the ordinary muscles of respiration, which are incapable of performing this function, and the patient necessarily dies of suffocation. The importance of recognizing these two factors in labored respiratory action, in view of the fact that one group of muscles is paralyzed by general anæsthesia, cannot be over-estimated.

In the choice of local anæsthesia cocaine is preferable to eucaine. The reasons for this preference are the following: Cocaine produces ischæmia, while eucaine produces hyperæmia. Inasmuch as hemorrhage, even slight, is of considerable significance in this operation, cocaine to that extent is preferable. The effect of cocaine is more prompt than that of eucaine, and, inasmuch as in this operation expedition is so often urgently demanded, the operation may be more quickly performed under the more quickly acting anæsthetic. And finally, the necessary dosage of either, when properly used, is insufficient to endanger life.

A one-tenth per cent. solution has been found of sufficient strength in my experience for inducing anæsthesia. It might be well to point out here that after the skin incision has been made, but few sensory nerve-terminals will be encountered in the section through the subcutaneous tissue until the structure of the trachea or the larynx has been reached. It is necessary to pause here and inject a new line of the anæsthetic. It is sometimes unnecessary to inject any anæsthetic in the dissection between the skin and the surface of the respiratory tube. After making the incision into the respiratory tube it is advisable to apply cocaine or eucaine upon the mucosa by means of a swab or spray for some distance above and below the opening, at least a distance sufficient to protect the area of possible operative procedures.

It is especially important, if the larynx can be reached from the point of opening, to cocainize its interior. This having been done, sudden collapse and death are quite impossible. This statement may seem rather positive, but any one having seen the emphatic experimental results and having tried this technique clinically, cannot say less. It matters not whether the operation is performed under general anæsthesia or under local, this precaution should be taken, inasmuch as it is true that, while general anæsthesia protects the patient against pain, it does not prevent the reflex inhibition, so that cocainization serves precisely as important a purpose under general anæsthesia as it does under local.

In the choice of general anæsthetics the patient may be more quietly and readily reduced under chloroform than under ether, and the tendency to the secretion of mucus is less. In these respects chloroform is preferable to ether. In another very important respect ether is preferable to chloroform,—namely, under chloroform anæsthesia there is a decidedly greater tendency towards respiratory failure and reflex inhibitory phenomena than under ether anæsthesia.

This observation on the comparative effect of chloroform and of ether is in full accord with work done by

Gaskill and Shore, Leonard Hill and Waller, whose conclusions, based upon a large number of experiments, show that chloroform exerts a much more toxic action upon the muscle-fibres of the heart than does ether, so that, in a given case of full chloroform anæsthesia, it is probable that the heart-muscle is more easily inhibited than under ether anæsthesia, producing, as it does, less direct toxic action upon the heart-muscle.

On the whole, it would probably be best to reduce the patient to surgical anæsthesia by means of chloroform, then continue the anæsthesia under ether, unless ample precaution against collapse be made by the local application of cocaine. Inasmuch as it is not the intention of this paper to deal with the general aspects of these questions, no comments will be made upon other points in the technique or treatment.

It has been frequently observed that if the foreign body is not lodged high up in the respiratory tract it is likely to be found at the bifurcation or in one of the bronchi, most frequently the left. There is, it seems to me, an additional explanation for this fact aside from that usually made. It is readily demonstrable that the expiratory efforts of the respiratory apparatus are more powerful than the inspiratory. It is necessarily true that an inspiratory force exerted upon a foreign body cannot exceed that of the atmospheric pressure rushing from an area of higher pressure to that of a lower,—that is to say, rushing into the chest during the formation of a partial vacuum therein. The expiratory force is limited only to the capacity of all the ordinary muscles of respiration plus that of all the extraordinary muscles of respiration, which collectively include large groups of muscles, many of which are the strongest muscles of the body.

It has been shown by experimental physiologists that the expiratory groups of muscles are more powerful than are the inspiratory. It follows, then, other things remaining the same, that a foreign body subjected to the play of these two factors, acting in opposite directions, must find a lodgement towards the side of the one which is weakest. Now, when the foreign body has dropped down so low as to become engaged in one of the bronchial tubes, the force of the respiratory efforts is only partially applied to this lodgement of the foreign body because of the want of resistance in the free bronchial tube; while, if the foreign body were in the trachea, the entire force of the expiratory efforts would be directed against the foreign body, and so would likely drive it onward into the larynx.

On purely theoretic grounds it would seem that if in a given case of lodgement of a foreign body in one bronchial tube, after a failure of respiratory efforts to dislodge it, the free bronchial tube were closed by an obturator quickly applied, so that the whole expiratory force might be brought to bear upon both bronchial tubes alike, thereby increasing to a great extent the expulsive force exerted upon the foreign body, such additional force might occasionally dislodge the foreign body that has resisted all other efforts. This, however, is a mere theoretic suggestion, not founded upon either direct experimental or clinical evidence.

In fixing the trachea with a hook, and, indeed, in any portion of the technique, it is of importance not to drag from above downward with too much force, else there may be produced an inhibitory effect upon the respiration, and, if the force be very severe, also upon the heart. This does not so much apply to traction from the lungs upward in the line of the tracheal axis. Lifting the trachea out of its bed forcibly is objectionable for the same reason that dragging down too forcibly upon the trachea should be avoided.

LARYNGOTOMY

Preliminary Remarks.—Surgeons have not infrequently encountered in performing a laryngotomy a sudden collapse at the moment when the incision was made through the larynx, and the margins of this incision were kept apart for the introduction of the tube. Even death has occurred not infrequently at this time. Such results have been more frequently encountered in hasty operations, and hasty operations for admitting air into the tract have usually been laryngotomies. In the history of tracheotomies, collapse, or death, at the particular stage referred to above, have been rarely observed.

Principles involved in the Technique.—The experimental and clinical evidence set forth in the preceding subject include all the principles involved in such results in laryngotomy. The larynx having been reached in the dissection, an incision is usually made boldly through it, and the margins of this incision are held apart, a procedure in which there is a mechanical irritation of the dangerous area in the larynx. This area, as has been shown experimentally, occupies the middle and upper portion of the larynx. It is well, then, if a laryngotomy must be made, to bear in mind this "inhibition" area in the larynx. It would add greatly to the safety of the operation if the incision were first made through the crico-thyroid space, then a swab of cocaine passed through this incision and applied to the laryngeal mucosa. This having been well done, no amount of manipulation could cause any reflex phenomena. Likewise, in the introduction of the tube in the high laryngotomy, there would be almost certain interference with the inhibition area and the production of the usual symptoms. Even though artificial respirations may be supplied, it must be remembered that the heart may be inhibited as well. The cocaine is an almost certain preventive of this cardiac inhibition, yet it would be safer to administer a hypodermic injection of atropine before beginning the operation.

Treatment of Reflex Phenomena in Laryngotomy.—What has been said on this subject under intubations may be said for the treatment of like conditions in this operation.

TRACHEOTOMY

Preliminary Remarks.—I am unaware of instances of sudden collapse or death in the technique for tracheotomy due to other causes than the obstruction for which the operation was performed,—that is to say, the collapse and death which occasionally follow intubation and laryngotomy do not follow the performance of tracheotomy.

Experimental.—Animals under surgical anæsthesia were subjected to the following experiments: The trachea was submitted to dilatation of different degrees, ranging from gentle dilatation to a sufficient force to rupture that organ, and in not a single instance was there noted any marked change in either the respiration or the circulation. All portions of the trachea from the larynx down were tested, and in all the cases the results were similar. The only variations observed were in some cases an increased respiratory rhythm with an increase in the amplitude of the respirations, and in some instances there was slight increase in the blood-pressure without any alteration in the character of the cardiac action. In several instances there was a slight decline in blood-pressure. It appeared that these several minor alterations in the blood-pressure and the respiration were due to mechanical stimulation of sensory tissue having a direct connection with the inhibitory apparatus of either the respiratory or the cardiac apparatus. From an experimental stand-point it would seem to be impossible to produce sudden death in any such manner as it may be produced in operations involving the larynx and certain portions of the pharynx.

The experiments upon the trachea were made in animals upon which experiments were being made for other purposes, and consequently they are not recorded separately.

Practical Application.—The result of these experiments is of practical importance mainly in pointing out the very great safety of operative procedures, so far as the immediate results are concerned, upon the trachea as compared with like operations upon the larynx,—that is to say, in considering the choice of tracheotomy or laryngotomy the principles involved in the two operations would be of great value in making a choice.

INTUBATION

Preliminary Remarks.—In performing intubations not infrequently cases of sudden collapse or death are encountered. These results appear with such suddenness as to give but little time for thought or action, and it would be a matter of great interest to those who perform such operations to know by what means such results are produced. In my own experience, in a series of sixty-five intubations, there were three such sudden deaths. While different theories have been proposed in explanation, none have, so far as I am aware, been founded on any considerable experimental evidence. The results are most frequently attributed to sudden suffocation from forcing down into the air-passage detached membranes. Others have said that the cases die of collapse, which is, of course, not an explanation. It has been asserted that the deaths are due to reflex phenomena from irritation, but the method of the production of these phenomena is not pointed out.

Protocols.—1. Healthy fox-terrier, weight twenty pounds, under full surgical ether anæsthesia, with the respiratory apparatus marking the blood-pressure taken in the carotid artery. After obtaining a control tracing, the larynx was dilated in a manner as nearly as possible like that of introducing a tube in intubation. The result of the careful introduction was merely a temporary inhibition of respirations. After a few feeble respiratory efforts, whose intervals were prolonged, normal respiratory rhythm was again established. However, when the tube was dragged forward with some force, so that its lower end pushed backward against the rima of the larynx and its upper end pressed forward against the rima glottidis, a more marked respiratory failure was produced, the carotid manometer executed long, sweeping strokes, and the blood-pressure during this time fell. On continuing the manipulation for some time a fair degree of tolerance was acquired, and the foregoing phenomena appeared in but a slight degree,—that is to say, the heart-strokes were a little longer and a little slower than normal, and the respirations were shorter, with a longer pause.

2. An old collie-dog, weight eighteen pounds, with the same arrangements for graphic records as in the preceding. A tube was introduced with gentleness into the larynx, then an attempt was made by exerting forcible manipulation upon the exterior of the larynx to force the tube out of the larynx into the mouth.

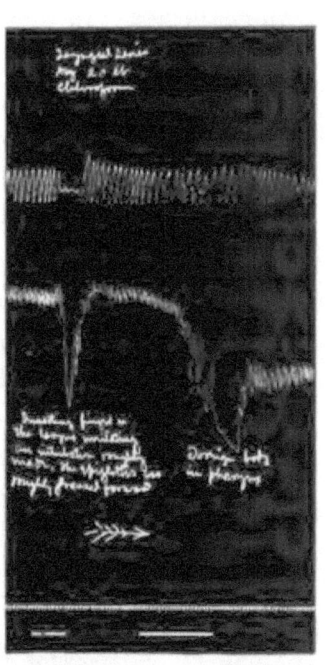

19. EXPERIMENT ON INTUBATION, ROUGHLY MADE, AND FOREIGN BODY IN THE PHARYNX.—The upper tracing represents the respiration, the next the blood-pressure, and the lower two the time and the signal respectively.

This rude manipulation produced an arrest of respiration with a marked fall in the blood-pressure, the heart-strokes being much slowed. On cessation of the manipulation, after the tube had been forced out, the respiratory and circulatory mechanisms resumed their normal action.

3. A mongrel cur, weight twenty pounds, with preliminary preparation as in the preceding case, was so placed upon the dog-board as to bring his mouth into a position convenient for the introduction of the fingers into the pharynx and the tilting the epiglottis forward with gentleness. In so doing there was but little disturbance in the respiratory and the circulatory tracings. The only alteration noted was in a single partial respiratory inhibition. The heart was not at all affected. Then, allowing a good control of tracings to be made, the fingers were again introduced, and with considerable roughness and force dragged forward the epiglottis, and with it the base of the tongue and larynx, imitating a vigorous effort to dislodge membrane or recover a tube in certain catastrophes, which occasionally attend the operation of intubation. Immediately upon so manipulating the parts there was a respiratory arrest and the heart was inhibited, causing a staggering, temporary fall in the blood-pressure and the appearance of "vagal" beats. On repeating such vigorous manipulations at short intervals a less marked effect on each repetition was observed, until, finally, when the animal had become quite exhausted, but little effect was produced.

4. With preliminary preparation as in the preceding case, the diaphragm was exposed by a median incision, and the stomach and the liver so withdrawn as to permit direct observations. The larynx was now subjected to severe manipulation by passing the finger, through an incision between the upper ring of the trachea and the cricoid, into the larynx and making pressure between this finger so placed and the thumb upon the exterior,

producing thereby inhibition of the respiration and the cardiac action as before. Observations upon the diaphragm during this time showed this organ to be in relaxation. The skin was then so removed from the chest as to expose the muscles of respiration, and the foregoing experiment again performed. Direct observation showed that the muscles were in a state of relaxation. No respiratory muscle was observed to be in a state of tetanic contraction. On section of both vagi the respiratory phenomena appeared as before, but the cardiac remained normal.

5. A shepherd-dog, twenty-five pounds in weight, preliminary preparations as in the preceding case. The various phenomena were again produced by like procedures. On painting the laryngeal mucosa with four per cent. solution of cocaine, then repeating the intralaryngeal manipulation, no inhibition of either the respiration or of the heart was observed.

6. A mongrel, weight twenty-one and one-half pounds, preparations as before. After securing control inhibitions by manipulations as in preceding experiments, one-one-hundredth of a grain of atropine was injected into the jugular vein. Almost immediately following this there was a rise in the blood-pressure and the manometer strokes were shortened, indicating the loss of "vagal" influence. Then, on repeating the procedures described in the foregoing experiments, the respirations were arrested as before, but the blood-pressure remained unaffected. On severing the superior laryngeal nerves, then repeating the laryngeal manipulations, no respiratory inhibition followed.

Summary of Experimental Evidence.—These experiments show that the reflex inhibition is due to efferent impulses set up by mechanic irritation of the terminals of the superior laryngeal nerves.

CLINICAL OBSERVATIONS

The Effect upon Respiration.—In almost every case of intubation there is at least a temporary inhibition of respiration in a manner and to a degree entirely comparable to that obtained by experiments on animals. The extent of this inhibition depends upon the following factors: In cases in which the laryngeal mucosa is completely protected by a closely adherent membrane, reflex inhibition of respiration is observed to be the least marked. In cases in which the membranes have disappeared, and the surface is raw and has been subjected to much irritation, like procedures produce a more marked reflex inhibition.

I have made these observations in a number of cases clinically. It need hardly be said that the extent of inhibition is also in direct ratio to the amount of mechanic irritation produced by the placing of the tube. The greater the irritation, the greater the inhibition. In a child poisoned with toxins, and with but little vitality, inhibition may not be more marked at its very onset, but recovery of the normal state does not so readily follow.

Effect upon the Cardiac Action.—In a number of instances in which direct observations upon the pulse were made during the introduction of the tube, an inhibitory effect upon the heart was occasionally observed. The cardiac inhibition is, however, not nearly so frequently observed as the respiratory, and what has been said as to the variations of the respiratory inhibition may be said equally of the cardiac.

Collapse and Death due to Inhibition.—From the experimental and the clinical evidence at hand it is evident that collapse and death in intubation are due to the reflex inhibition of the respiration and of the heart, from mechanic stimulation of the superior laryngeal nerves in the manipulation of the tube. The onset of the symptoms is sudden. The patient at once becomes limp and lifeless, the face becomes livid, the pulse has disappeared from the wrist, respirations have ceased, and the child is dead.

In cases less severe the child suddenly goes into a collapse characterized by its becoming suddenly relaxed in the midst of a struggle while the tube is being placed, the respiration ceasing, and the pulse not being perceptible at the wrist. However, after a brief pause, respiration slowly resumes, the pulse may be felt, weak and slow, but gradually growing stronger, and in a few minutes respiration and the cardiac action are again restored. The most marked characteristic is the absolute suddenness of the cessation of the respiratory and the cardiac action.

Differential Diagnosis between Obstruction from Membranes pushed down and Collapse from Reflex Inhibition.—In asphyxia from pushing down the membranes the characteristic symptoms are the deep cyanosis, slow but full pulse, and continued respiratory efforts for a short time. In a case of pure asphyxia from obstruction, respiratory efforts do not suddenly cease. In collapse from

PERIMENTAL RESEARCH INTO THE

reflex inhibition the pulse disappears from the wrist, the heart-beats, if observable at all, are very slow and weak. In asphyxia the heart-beats grow temporarily actually stronger. In reflex inhibition death may instantly ensue. In asphyxia it cannot.

In asphyxia from obstructing membranes, cyanosis is much more marked than in reflex inhibition. In reflex inhibition, when very profound, there may be a sudden pallor on account of the total sudden failure of the circulation. There is an abundance of experimental evidence, and physiologists are agreed that the blood-pressure in asphyxia rises, that the heart-beats become stronger and slower, and there is clinical evidence entirely in accord with that of the experimental physiologists. The quick perception of the difference between these two conditions is of the utmost importance in the conduct of the case, as will be more fully pointed out directly.

Prevention of Collapse from Reflex Inhibition.—So far as the heart is concerned reflex inhibition may be wholly prevented by a preliminary hypodermic injection of atropine. This drug, as is well known, paralyzes the nerve-endings of the vagi in the heart, thereby protecting this organ against reflex inhibitory impulses. Or the local application of cocaine upon the laryngeal mucosa may prevent not only the reflex inhibition of the heart, but of the respiration as well.

The Treatment of Collapse from Reflex Inhibition.—It is now my practice so to arrange the clothing of the patient, before performing the operation of intubation, that ready access to the chest and the abdomen is provided, and that a towel and a basin of cold water are at hand.

On the appearance of collapse the patient is inclined, head downward, and artificial respiration maintained. In the mean time the chest and abdomen are smartly struck with a cold, wet towel. Immediately after this slapping, the surface is dried and the slapping again repeated. Cold water smartly applied to the surface of the skin is one of the most powerful respiratory stimulants, and an inspiratory effort is always first brought out on such treatment. Further than this there is little to be done. There is no time for medication. It will be at once seen how important it is to recognize the difference between collapse produced in such manner and asphyxia from forcing down membrane. Suppose in a given case of collapse from reflex inhibition it was believed that membrane had been forced down, the tube having been placed in the larynx, hasty efforts would be immediately directed towards relieving the respiratory tract of this obstruction. Such efforts would consist naturally of an attempt hastily to remove the tube, and the efforts at relief would by further stimulation of this inhibition area in the larynx cause an increase in the very condition the operator was seeking to relieve. And, furthermore, the blood-pressure having sunk almost to zero, the patient, instead of being inclined head downward, may be kept in the sitting posture while attempts at relieving the obstruction are being made. And, finally, it is doubtful, in consideration of the extraordinary expulsive power of the respiratory apparatus, whether a membrane may be detached and so firmly lodged below a tube as to cause an obstruction after removal of the tube.

If clinicians will take a retrospect of the cases in which

such an obstruction was supposed to exist, they will probably be enabled to recall no great respiratory efforts on the part of their patients. In the presence of so appalling a catastrophe, calm observations are by no means, as a rule, made, and if every case were approached with the deliberate intention of making correct observations, there would probably not be a mistake in confusing obstruction with reflex inhibition, and it would be found that nearly all the cases of so-called obstruction are cases of reflex inhibition. In one case occurring in my own practice, in which the child went into collapse while introducing a tube in a second intubation, the child was handed to one assistant, another kept up artificial respiration, while I made clinical observations of the phenomena. Placing my ear to its chest, I heard faint, slow heart-beats, soon becoming louder and more rapid as the child rapidly regained consciousness. This case was a repetition in its every phase clinically of the inhibition phenomena experimentally produced.

EXPERIMENTAL RESEARCH INTO THE CAUSE OF CERTAIN PHENOMENA ATTENDING CONSIDERABLE TRACTION ON THE TONGUE

Preliminary Remarks.—On several occasions I have observed an immediate cyanosis and very considerable collapse attending vigorous traction on the tongue on the part of an over-zealous anæsthetizer, in his efforts to relieve the patient from the obstruction due to the recession of the tongue upon the laryngeal opening. While before this vigorous traction there were respiratory efforts (though futile, on account of the mechanical obstruction alluded to), during and immediately following the traction there was complete cessation of respiration for a brief interval, then slow and shallow respiratory efforts, which finally and gradually became vigorous again. The pulse meanwhile was weak and slow; in short, the patients were in temporary collapse. It was a matter of doubt as to whether or not the phenomena described were due to the anæsthetic.

Protocols.—1. On a twenty-pound shepherd-dog, under ether anæsthesia, with initial blood-pressure at one hundred and twenty millimetres, the following experiments were performed: The upper jaw was fixed firmly upon the dog-board, the lower jaw was held widely open, the tongue was grasped with three Well's forceps; this produced no effect upon either respiration or the blood-pressure. Gentle traction was then made in the line of the axis of the tongue, during which the blood-pressure and the respiratory action remained unchanged; then, suddenly applying forcible traction, the heart was temporarily arrested, producing a staggering fall in the blood-pressure; following this temporary inhibition of the heart, there were slow, vagal strokes, gradually returning to the normal action. In the mean time the blood-pressure rose to the level it was before, simultaneously with the arrest of the heart, the respirations were completely stopped during the application of the severe traction. On releasing the tongue, the respirations were again resumed, at first more slowly, finally in the normal manner. The lower jaw was then sawn in the median line and the lateral halves were separated as far as possible, producing thereby a marked fall in the blood-pressure, the heart showing distinct vagal influence; the respirations meanwhile, after suffering a temporary arrest, resumed slowly their normal action. The effects were similar to those produced by the severe traction on the tongue, though less in degree. The first experiment of traction upon the tongue was twice repeated, each time yielding practically the same results.

2. On a Newfoundland dog weighing forty-five pounds, under ether anæsthesia, with an initial blood-pressure at one hundred and fifty millimetres, the following experiments were performed: The tongue was drawn out and securely tied with a heavy cord near the frænum; the blood-pressure and respiration meanwhile remained unaltered. With the aid of this string and three Well's forceps, the tongue was gradually subjected to a traction along the line of its own axis. In the first part of the traction no effects were observed. On the application of greater force the heart was very much slowed, the blood-pressure suffered a slight decline, and the respirations were greatly diminished in the amplitude of their stroke and in their frequency; on applying a still greater traction, the heart was temporarily arrested, as was the respiration also; then the lower jaw was opened with great force as widely as possible, and finally cardiac and respiratory effects similar to those observed in the preceding case, though to a less degree, were produced. The traction phenomena were twice produced by repeating the technique above described.

3. On a twenty-pound fox-terrier, with an initial blood-pressure of one hundred and forty millimetres, under chloroform anæsthesia, the following experiments were performed: The tongue was drawn out in the manner

previously described, then suddenly drawn across the angle of the mouth, thereby producing complete arrest of the heart's action as well as of respiration, with a consequent collapse in the blood-pressure almost to the abscissa line. This manœuvre was repeated a number of times in rapid succession, each time producing similar effects upon the cardiac action and the respiration, finally resulting in very great collapse, from which the animal did not recover.

4. On a fat poodle weighing twelve pounds the preceding technique was carried out as nearly as possible in every detail, with results practically the same.

5. On a twenty-four-pound mongrel under chloroform anæsthesia, with initial blood-pressure of one hundred and sixty millimetres, the various kinds of traction were applied, producing a most profound collapse by temporarily inhibiting both the heart and the respiration. A piece of gauze was then thrust down very roughly into the pharynx, in imitation of clearing out the throat during the administration of an anæsthetic. This produced a temporary arrest of respiration, but no effect upon the heart's action.

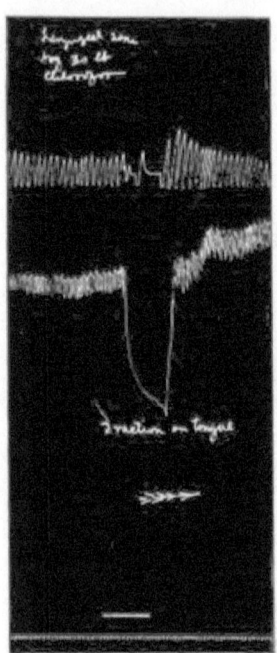

21. TRACTION ON THE TONGUE.— The upper tracing represents the respiration, the next the blood-pressure. Note the temporary inhibition of both during forcible traction. This was the most marked effect produced in any of the experiments.

6. On a fox-terrier weighing twenty-two pounds, under chloroform anæsthesia, the following experiments were performed: The tongue was subjected to traction, as in the preceding experiments, producing the usual phenomena, after which one-one-hundredth of a grain of atropine was injected into the jugular vein. This occasioned an immediate rise in the blood-pressure and an increase in the frequency of the heart-beats, indicating the release of the heart from the influence of the vagi. The left vagus was then exposed and the efficiency of the atropine tested by applying faradic stimulation; the stimulation did not affect the cardiac action, which proved the efficiency of the atropine. The tongue was then subjected to traction in the same manner, and, as nearly as possible, to the same degree as in the preceding experiments. Although this traction was repeated several times in various directions, instead of a fall in the blood-pressure there was produced a temporary rise. The respirations, however, were arrested as in the previous experiments. The lower jaw was then severed, and the lateral halves were widely separated, producing a rise in blood-pressure.

7. On a bull-dog weighing forty-eight pounds, under chloroform anæsthesia, with initial blood-pressure at one hundred and fifty millimetres, the following experiments were performed: The anæsthesia, during the preparation of the animal, was too profound, and artificial respirations became necessary. Before the animal was completely anæsthetized the jaws were held apart and a severe traction across the angle of the mouth was made upon the tongue. Although the respirations were rapid at the time it was made, they wholly ceased, until the animal became

very cyanotic under the influence of the powerful traction. The heart was simultaneously arrested, and the blood-pressure collapsed. A physiological dose of atropine, with the same proof of efficiency as in the preceding case, was injected. The tongue was again subjected to a similar experiment, but the traction produced no fall in the blood-pressure, although the respirations were effected as in the preceding case. This was several times repeated with similar results.

8. On an animal weighing thirty pounds, with initial blood-pressure at one hundred and sixty-five millimetres, a control was made, whereby the respiratory and the circulatory phenomena were produced as usual. Then both vagi were severed, and a similar procedure repeated. But in this case, as in the atropine experiment, the cardiac symptoms did not appear, although the respiratory arrest was as striking as before. This was repeated several times with practically the same results. The jaw was severed and separated, as in the previous experiment, producing thereby essentially similar results.

9. On a thirty-five-pound mongrel, under chloroform anæsthesia, with initial blood-pressure of one hundred and seventy millimetres, the following experiments were performed: A control was first secured in the usual way, producing the ordinary respiratory and circulatory phenomena. The superior laryngeal nerves were then both severed, and the preceding experiment repeated. The effect upon the respiration was wholly prevented, but the blood-pressure fell as before, the heart exhibiting strokes similar to those observed in the preceding case. Then the glossopharyngeal nerves were both severed, and the same

experiment of traction upon the tongue repeated. The respirations were not at all affected, but the heart was inhibited, and the blood-pressure fell, as in the preceding case. Finally, the glossopharyngeal nerves were both severed near their exit from the cranium. The tongue was then subjected to traction, as before, with the result of arresting the heart, without in the least influencing the respiration. Finally, both vagi were severed, and similar experiments were performed. Neither the cardiac nor the respiratory action was now in the slightest degree affected.

10. In this experiment the technique and plan described in the ninth were carried out as nearly as possible, resulting in essentially the same observations.

In a number of other experiments the effects of traction upon the tongue were tried, and in a few cases slight traction produced a slight rise in the blood-pressure. In one instance, no effect was produced on either the respiration or the heart.

Summary of Experimental Evidence.—When any effect was observed from traction of moderate degree, it was usually a slight rise in blood-pressure, though generally such traction produced no effect.

In most animals following considerable traction, more especially when the force was so applied as to drag the tongue out at the angle of the mouth, there were varying degrees of fall in the blood-pressure, ranging from a complete temporary cardiac arrest, with an enormous fall in pressure quite to the abscissa line in several experiments, to a series of slowed beats with but slight actual fall of the mean pressure.

Then, again, there would be an intermission of one or

more beats with rapid recovery of the lost pressure. In some experiments no effect was produced on applying the traction.

The effect on respiration was to cause in many instances total temporary arrest; in a lesser number there was a decrease in the rapidity as well as in the amplitude.

Occasionally no respiratory alterations were produced.

Preliminary injections of a physiologic dose of atropine prevented the cardiac phenomena, but not the respiratory. Like results were obtained by preliminary section of the vagi.

A sufficient traction on the tongue usually causes inhibition of the respiratory and the cardiac action, the intensity of the inhibition ranging from a slowed action to complete arrest.

The effect upon the blood-pressure and the respiration was at first believed to be wholly due to mechanic stimulation of the superior laryngeal nerves. However, when these nerves were severed the respiratory phenomena were entirely prevented, but the heart exhibited the same inhibition as before.

Section of the vagi prevented this effect upon the heart; the effect was, therefore, through the vagus. All the nerves supplying this region were in turn severed, but the effect upon the heart was not thereby prevented. We therefore concluded that the action upon the heart was due, at least in part, to a stimulation of the vagus by dragging upon the structures in such close anatomic relation with it as to produce thereby a mechanical stimulation upon them.

On making a dissection of this portion of the neck and

observing the effects of traction on the tongue, such results were clearly seen to have occurred. Mechanic stimulation by dragging directly upon the vagus may produce an effect upon the heart quite similar to the effects described in the protocols. The effect upon the respiration was due to a mechanic stimulation of the superior laryngeal nerves.

Practical Applications.—These results are of great practical importance in the routine work of anæsthesia. A thorough appreciation of the inhibitory phenomena produced by excessive traction, would enable the anæsthetizer to prevent their appearance and the surgeon to more intelligently meet them when they do occur.

THE END

www.ingramcontent.com/pod-product-compliance
Lightning Source LLC
Chambersburg PA
CBHW030406170426
43202CB00010B/1512